327

Globalisation

Simon Oakes

WITHDRAWN

Advanced Topic*Master*

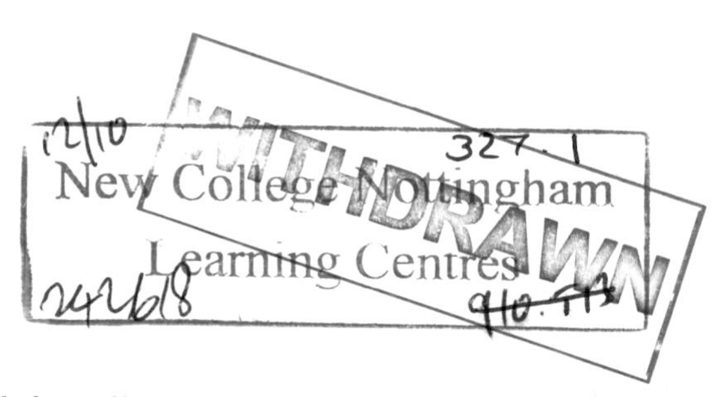

Philip Allan Updates, an imprint of Hodder Education, an Hachette UK company, Market Place, Deddington, Oxfordshire OX15 0SE

Orders
Bookpoint Ltd, 130 Milton Park, Abingdon, Oxfordshire OX14 4SB
tel: 01235 827827
fax: 01235 400401
e-mail: education@bookpoint.co.uk
Lines are open 9.00 a.m.–5.00 p.m., Monday to Saturday, with a 24-hour message answering service. You can also order through the Philip Allan Updates website: www.philipallan.co.uk

© Philip Allan Updates 2010

ISBN 978-1-84489-640-0

First printed 2010
Impression number 5 4 3 2
Year 2014 2013 2012 2011 2010

All rights reserved; no part of this publication may be reproduced, stored in a retrieval system, or transmitted, in any form or by any means, electronic, mechanical, photocopying, recording or otherwise without either the prior written permission of Philip Allan Updates or a licence permitting restricted copying in the United Kingdom issued by the Copyright Licensing Agency Ltd, Saffron House, 6–10 Kirby Street, London EC1N 8TS.

Printed in Spain

Hachette UK's policy is to use papers that are natural, renewable and recyclable products and made from wood grown in sustainable forests. The logging and manufacturing processes are expected to conform to the environmental regulations of the country of origin

P01642

Contents

Introduction

This book provides a detailed and up-to-date review of globalisation for students of geography at advanced level. It is aimed primarily at sixth-form students following A-level and International Baccalaureate courses, but the content of the book may also be useful for first-year undergraduates.

Globalisation is a set of processes whose outcome (measured in terms of people's interconnectedness and their level of immersion in global economic and social networks) can be calculated through the use of various indices. The data show that globalisation has accelerated markedly, most particularly between the 1970s and the early years of the twenty-first century. Economic liberalisation and internationalism of communications and transport networks have enabled new technologies to operate across national borders, creating the sense of a 'shrinking world' — a key geographical outcome of globalisation.

How have increased global interactions impacted on the lives of people and places? Expert opinion remains divided with regard to the net effect of globalisation on poverty and quality of life in what used to be called 'the global south'. Enormous social and cultural transformations are occurring on a global scale, linked with the transmission and diffusion of cultural traits by transnational corporations (TNCs) and international migrants. In some contexts, these processes have resulted in the lessening of cultural diversity, while in others we find increasing local diversity and a new hybridity. Globalisation is also associated with environmental degradation at varying scales, including the planetary crisis of climate change.

The political response to globalisation has been complex. The challenges and opportunities brought by increasing global interactions have prompted a range of responses at different scales. These include accelerated growth of supranational organisations and an associated loss of autonomous power for some sovereign states, notably those belonging to the European Union. A wider range of political challenges to globalisation, including resistance, protests, opposition or 'opting-out', show that global interactions continue to be contested and negotiated. Recent world events have also shown that globalisation may comprise a less stable set of internal processes than many people imagined during its 'golden age' in the 1990s and early 'noughties'. Many of the world's leading economies, including the UK and the USA, spent much of 2009 in recession. During the global credit crunch of 2008–09, an unprecedented decoupling of global linkages and flows took place. Trade and investment flows slowed, while many international migrants headed home as overseas work opportunities contracted.

Students can use this book in several ways. The material provides a general account of the growth, impact and challenges of globalisation, reflected in the three main divisions of the book. The text should be read to help you consolidate your understanding as each new topic is covered in class. Specific areas of the text can be used to complete essays and other assignments. The diagrams, tables and photographs should be studied carefully and used to illustrate or extend the concepts and facts covered by the main text. General discussion and explanation is backed up by contemporary case studies, which can be used to support assignments and written work in your examination. Finally, activities interspersed throughout the text aim to deepen your knowledge and critical understanding, to develop evaluative skills, including the statistical analysis of data, and to encourage you to investigate these exciting topics further and develop an enduring interest in global current affairs and connectivity .

Simon Oakes

1 The nature of globalisation

Modern globalisation

Modern globalisation is not without history. It is the continuation of a far older, and ongoing, economic and political project of global trade and empire-building. Economies have been interdependent to some extent since the time of the world's first great civilisations, such as ancient Egypt, Babylon and Rome. More recently, during the nineteenth century, the British empire gifted the British people and the English language with a global sphere of influence. There is nothing novel in the global power-play and ambition of strong individuals, nations and businesses.

However, globalisation today differs markedly in many respects from earlier worldwide enterprises. Flows of money, materials, ideas, information and people have always linked together different places, societies and environments; but the *nature* of these flows began to change in three important ways during the later decades of the twentieth century:

- Connections grew longer and more extensive, linking greatly increased numbers of localities together. Many more places and people are now economically and socially interdependent than in the past. Enormous and complex networks of production and consumption evolved to the point where UK firms like Tesco routinely open stores in faraway southeast Asia.
- Connections began to run deeper into ordinary people's lives. In the early decades of the twentieth century, only a privileged aristocracy lived 'global' lives. Dukes and earls might visit southern Africa on safari, but working-class people rarely left their own locality. Theirs was a provincial existence, consuming predominantly local food and clothing. In contrast, cheap flights now bring global travel in reach of most citizens living in richer nations and are no longer the preserve of an economic elite. Likewise, even people in the slums of Rio or the socially disadvantaged areas of London or Paris, for example, are now consumers in a 'global village', surviving on cheap food and clothes that often have distant origins.

- Faster connections are now operating in real time. Technical changes have enabled the world to experience **time–space compression**. Distance between continents has become measured in mere hours of flight time. Air travel, the telephone, internet access and containerised shipping are among the crucial new technologies that help coordinate economic, political, cultural and sporting activities taking place simultaneously in different parts of the world. In virtual environments such as Facebook, truly globalised communities work and play in perfect synchronicity.

The *impacts* of globalisation have also increased over time, especially during the last 30 years. First, the full implications of climate change — causally linked with the runaway carbon emissions of globalised mass consumption — have only recently become apparent and the sustainability of large numbers of terrestrial and marine ecosystems is increasingly jeopardised. The economic transformation of billions of people's lives — for better or worse — is a second key impact of globalisation. In 1960, a large part of the world's population remained disconnected from any global trade flows and instead engaged in local, rural subsistence activities. Global agribusiness has now largely transformed this way of life, modernising farming systems and enlisting peasant farmers into a salaried cash-cropping workforce. Those left without land or employment often migrate to cities in search of work in the factories and offices of yet more global employers — the urban-based **transnational corporations** (TNCs). Of the 20 million people who live in the São Paulo area today, half are jobseeking migrants not born there.

Are these enormous planet-wide changes inevitable or permanent? The **global credit crunch** — a financial crisis triggered in 2008 by the collapse of many major American and European banks — led some commentators to speculate that globalisation was in serious jeopardy (see Chapter 9). Mere months later, at the start of 2009, an estimated 20 million migrant workers in China had quit the cities and returned temporarily to their countryside homes. The global slowdown had precipitated a fall in worldwide demand for manufactured goods, bringing a sudden wave of deindustrialisation to China's manufacturing boomtowns of the 1990s such as Chongqing (growth has since resumed).

However, globalisation (and the world economy that preceded it) has never grown in a linear and uninterrupted way. Previous phases of global capitalist expansion were halted by the world wars of 1914–18 and 1939–45 and by international recession during the 1880s and 1930s. Postwar globalisation has so far experienced several significant shocks, of which the global credit crunch is just the latest — though perhaps the most serious (Figure 1.1). If history is any guide, periods of stagnation and decline are to be expected; but the underlying long-term trend is one of continued growth, for both interconnectivity and total global wealth.

Figure 1.1 World economic growth, 1971–2009, and annual percentage increase in gross world product (GWP)

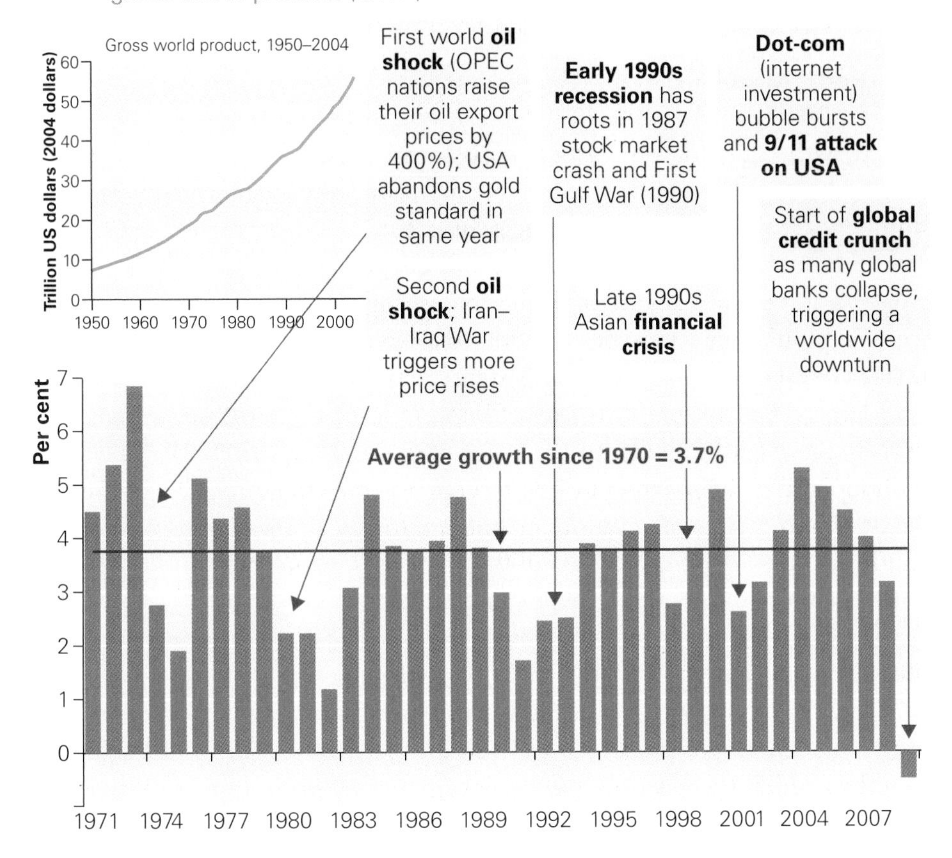

A multi-strand process

So far, we have discussed globalisation in terms of changing connectivity between places. However, there are many more ways of attempting to define it (Figure 1.2). The sheer enormity and complexity of the processes and outcomes involved means there is much variety in how globalisation is described and accounted for by people working in different professions. Geographers, economists, historians and sociologists all have their preferred definitions.

Figure 1.2 Defining globalisation

A rapid and huge increase in the amount of economic activity taking place across national boundaries has had an enormous impact on the lives of workers and their communities everywhere. The current form of globalisation, with the international rules and policies that underpin it, has brought poverty and hardship to millions of workers, particularly those in developing and transition countries. *UK Trade Union Congress*

Globalisation is a process enabling financial and investment markets to operate internationally, largely as a result of deregulation and improved communications. *Collins dictionary*

The term 'globalization' refers to the increasing integration of economies around the world, particularly through the movement of goods, services, and capital across borders. It refers to an extension beyond national borders of the same market forces that have operated for centuries at all levels of human economic activity — village markets, urban industries, or financial centres. There are also broader cultural, political, and environmental dimensions of globalization. *IMF*

Globalisation

Globalization can be conceived as a set of processes which embodies a transformation in the spatial organization of social relations and transactions, expressed in transcontinental or interregional flows and networks of activity, interaction and power. *Held and McGrew (Globalization Theory)*

The expansion of global linkages, organization of social life on global scale, and growth of global consciousness, hence consolidation of world society. *Frank Lechner (The Globalization Reader)*

It might mean sitting in your living room in Estonia while communicating with a friend in Zimbabwe. It might mean taking a Bollywood dance class in London. Or it might be symbolized in eating Ecuadorian bananas in the European Union. *World Bank (for schools)*

The world's economies have developed ever-closer links since 1950, in trade, investment and production. Known as globalisation, this process is not new, but its pace and scope has accelerated in recent years, to embrace more industries and more countries. The changes have been driven by liberalisation of trade and finance, how companies work, and improvements to transport and communications. *BBC*

Activity 1

Study Figure 1.2. To what extent do the definitions appear similar, even if they are differently worded? Or do the various writers, depending on their own vested interests, define globalisation in significantly different ways — suggesting that the meaning of the word is contested? Construct a table with two columns, headed 'similarities' and 'differences', to summarise your findings.

Some analysts subdivide globalisation into a series of constituent issues and tackle each separately. Globalisation thus becomes a multi-strand process, whose many linked themes are intertwined like the strands of a rope. For instance, economic globalisation, political globalisation, social globalisation and cultural globalisation can be identified as specific but interconnected strands (Figure 1.3).

Figure 1.3 Four strands of globalisation

Economic globalisation
- *The growth of transnational corporations (TNCs)* accelerates cross-border exchanges of raw materials, components, finished manufactured goods, shares, portfolio investment and purchasing
- *Information and communications technology (ICT)* supports the growth of complex spatial divisions of labour for firms and a more international economy
- *Internet and the World Wide Web* has allowed extensive networks of consumption to develop (e.g. online purchasing using eBay or Amazon)

Social globalisation
- *International immigration* has created extensive family networks that cross national borders — world city-societies become multi-ethnic and pluralistic
- *Global improvements in education and health* can be seen over time, with rising world life expectancy and literacy levels, although the changes are by no means uniform or universal
- *Social interconnectivity* has grown over time thanks to the spread of 'universal' connections such as mobile phones, the internet and e-mail

Political globalisation
- *The growth of trading blocs (e.g. EU, NAFTA)* allows TNCs to merge and make acquisitions of firms in neighbouring countries, while reduced trade restrictions and tariffs help markets to grow
- G7/G8 and G20 groups of countries regularly meet to discuss global concerns such as the economy and environment (climate change and the credit crunch being two major issues of today)
- *The World Bank, the IMF and the WTO* work internationally to harmonise national economies

Cultural globalisation
- *'Successful' Western cultural traits come to dominate* in some territories, e.g. the 'Americanisation' or 'McDonaldisation' of tastes and fashion
- *Glocalisation and hybridisation* are a more complex outcome that takes place as old local cultures merge and meld with globalising influences
- *The circulation of ideas and information* has accelerated thanks to 24-hour reporting; people also keep in touch using virtual spaces such as Facebook and Twitter

Origins and growth of globalisation

By the 1700s, extensive Asian and Atlantic trading zones were already in existence (Figure 1.4). The former linked China, Japan, southeast, south and central Asia and Arabia. The latter connected Europe with the Americas and Africa (its roots lie in Portuguese expansion into western Africa from the early 1400s). Flows of slaves and raw materials soon began to bind together the economic destinies of colonising and colonised Atlantic nations. Globalisation of major European languages was also well under way, initiating a mass extinction of indigenous local languages that continues today (see Chapter 5, pages 62–63). The most important technical development during the earlier phase of world-scale economic and cultural network-building was the telegraph, followed quickly by the telephone. By the late 1800s, people in far-distant places were communicating with one another in real time.

Figure 1.4 Global trading zones in the sixteenth to eighteenth centuries

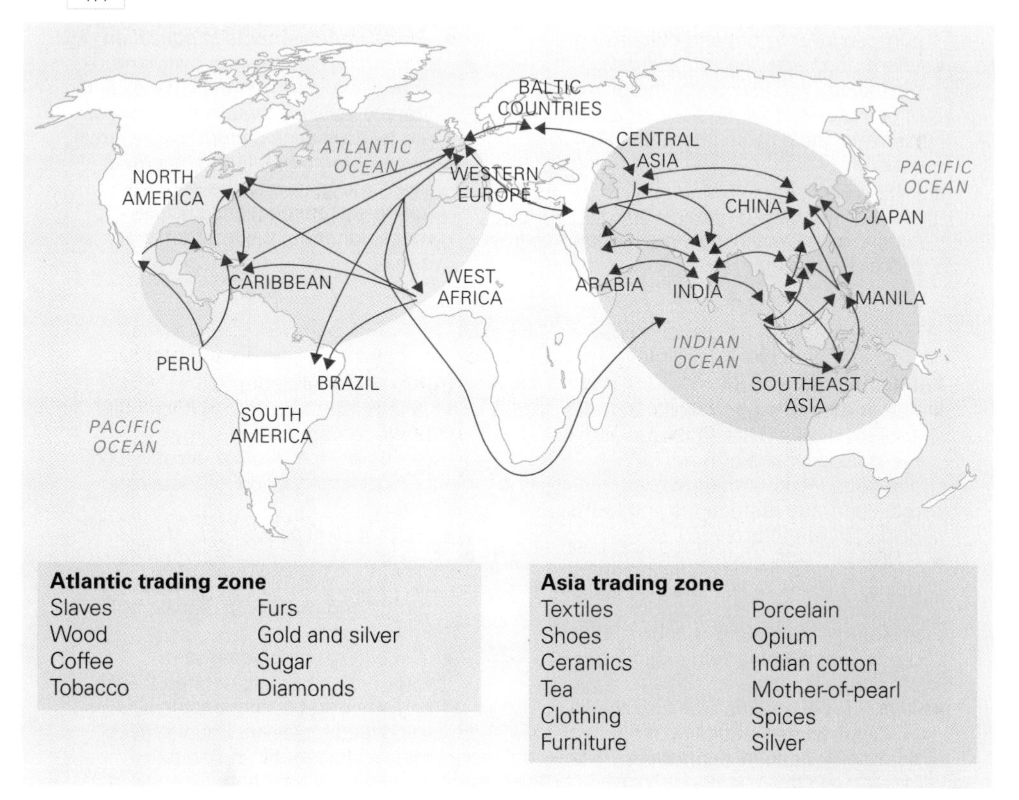

Activity 2

Study Figure 1.4.

(a) Describe the key features of world trade as shown in Figure 1.4.

(b) Using your own knowledge, describe and explain key ways in which trading patterns today may differ from these older ones. Possible factors to consider (among many others) include changes in the distribution of world population; new political alliances; technological developments (including canal-building).

Figure 1.5 Postwar timeline for economic globalisation

1944–45 Establishment of the World Bank, International Monetary Fund (IMF) and origin of the World Trade Organization (WTO). The postwar Bretton Woods conference provides the blueprint for a free-market non-protectionist world economy where aid, loans and other assistance become available for countries prepared to follow a set of global financial guidelines written by powerful nations.

1960s Heavy industry in the developed economies of Europe and America is increasingly threatened by rising production in southeast Asia, including Japan and the emerging Asian Tiger economies, notably South Korea. Unionised labour costs push up the price of production for Western shipbuilding, electronics and textiles. Most older economies enter a period of falling profitability for industry.

1973 The first OPEC oil crisis pushes Western industry over the edge. Rising fuel costs trigger a 'crisis of capitalism' for Europe and America, whose firms begin to step up outsourcing and offshoring of their production to low-labour-cost nations. Meanwhile, soaring petrodollar profits for Middle East OPEC nations signal that the United Arab Emirates and Saudi Arabia are on their way to becoming new global hubs.

1980s Financial deregulation in major economies like the UK and USA brings a fresh wave of globalisation, this time involving financial services, share-dealing and portfolio investment (by 2008, financial markets have an inflated value more than twice the size of actual world GDP!). The collapse of the Soviet Union in 1989 significantly alters the global geopolitical map, leaving the USA as the only 'superpower'.

1990s Landmark decisions by India (1991) and China (1978) to open up their economies bring further change to the global political map. Established powers strengthen their regional trading alliances, including the EU (1993) and North American Free Trade Association (1994). The late-1990s Asian financial crisis is an early warning of the risks brought by loosely regulated free-market global capitalism.

2000s Global credit crunch (2008). Major flaws in the globalised banking sector emerge as unsecured loans totalling trillions of dollars undermine leading banks. This brings a negative multiplier effect causing high-street store closures to the UK. In an interconnected world, growth slows for the first time in two decades for China and India, two great outsourcing nations and new emerging superpowers.

During the immediate postwar period, the coincidence of several key political, financial and technological influences further began to mould the world's nations into a single, global economic unit (Figure 1.5). 'First World' governments purposely nurtured global trade through various means, and for varying reasons:

- New guiding principles for governments seeking financial support were established by the World Bank and International Monetary Fund (IMF), twin organisations set up in Washington with strong US backing. Their initial

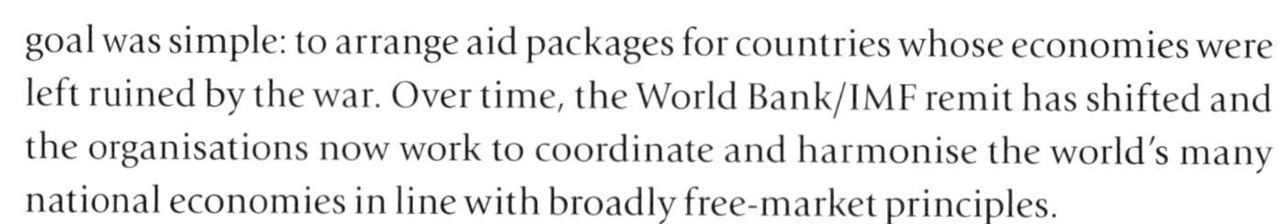

goal was simple: to arrange aid packages for countries whose economies were left ruined by the war. Over time, the World Bank/IMF remit has shifted and the organisations now work to coordinate and harmonise the world's many national economies in line with broadly free-market principles.

- Under the free-trade umbrella, countries were encouraged to exploit their own comparative advantages and build economies of scale for export-led industries, the most successful of which have grown into enormous global corporations (such as South Korea's Samsung).
- The key guiding principle was to avoid a return to conditions that prevailed during the Great Depression of the 1930s, when a global economic downturn led to free trade becoming replaced by **protectionism**. Nations had blocked foreign imports with tariffs, damaging export markets for other countries and resulting in a vicious downward spiral of economic output for all major players.
- Another important consideration for the USA and western Europe during the 1950s and 1960s was to offer a clear development path for Asian, African and Latin American nations which might otherwise be attracted to the Soviet brand of communism.

Today, receipt of IMF financial loans and development packages by emerging economies remains conditional on the beneficiaries opening themselves up to foreign investment by TNCs seeking labour supplies, markets and natural resources. Changes in the nature of TNCs and the regulations that bind them also feature in the story of globalisation. Financial deregulation for many European countries and the USA has led to banks and finance companies globalising rapidly through mergers and foreign acquisitions.

The World Trade Organization (WTO), formed in 1995 as a successor to the General Agreement on Trades and Tariffs (GATT), also advocates trade liberalisation, especially for manufactured goods such as textiles and electronics. It encourages countries to abandon protectionist attitudes in favour of non-restrictive importing and exporting of manufactured goods, unfettered by prohibitive taxes and tariffs. However, the WTO has been less successful in persuading the richest countries to adopt a similar 'level playing field' for agricultural imports and exports: EU nations and the USA continue to support and subsidise their own farmers, using a range of protective measures. The result is the often-heard criticism that the interests of developed nations are repeatedly favoured by the WTO above those of poorer countries which still trade mostly in primary food products.

Figure 1.6 Indonesia's globalisation timeline

Globalisation phases	Factors
 **Phase 1: Colonised by the Dutch** 1670–1945 For nearly 300 years, Indonesia experiences an early form of globalisation as an exploited and dependent European colony. It is also invaded by the Japanese during the Second World War. 	• **Raw materials** drew the Dutch to Indonesia. Flows of valuable tin, copper, timber, rubber and gold made Indonesia's islands an important prize.
Phase 2: Independence and anti-Western stance 1946–64 After 4 years of guerilla warfare, new leader General Sukarno takes power, with a strong anti-Western stance. The message 'Go to hell with your aid' is sent to the USA and the IMF. Sukarno courts the communist Soviet Union. 	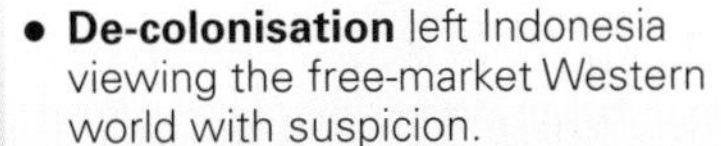 • **De-colonisation** left Indonesia viewing the free-market Western world with suspicion. • **The Cold War** (when the USA and the Soviet Union were opposing world superpowers) was a major influence on globalisation of Indonesia.
Phase 3: Regime change and pro-Western stance 1965–67 Backed by Washington, General Suharto seizes power from General Sukarno during a period that sees hundreds of thousands of suspected Indonesian communists murdered. 1968 Suharto's new regime opens up Indonesia's economy. American and European TNCs meet with the Suharto government in Switzerland and design new attractive economic policies for free-market investors coming to Indonesia. 	• **IMF infrastructure loans** Western TNC branch plants arrived in Indonesia as soon as its roads, power supplies and ports began to modernise. • **Legal changes** The legal framework for the export processing zone in Jakarta gave TNCs exactly what they needed — a low-tax haven for sweatshop manufacturing, making the most of low labour costs.
Phase 4: Economic collapse and recovery 2000+ Poor publicity results in many foreign investors like *Gap Inc.* improving conditions for workers. After nearly 40 years as a major global player, Indonesai is the world's 16th-largest economy. However, progress still needs to be made for Indonesia's poor (the country is ranked just 115th when GDP per capita is measured).	• **Anti-globalisation** protests have put the spotlight on Indonesia's sweatshops and the tough line taken by the Suharto regime against trade union campaigners, many of whom were imprisoned. Half the population live on less than two US dollars a day. • **G20 membership** has made Indonesia a major world political player, adding another dimension of globalisation to the nation's profile.

Throughout the postwar period, a consistent feature of globalisation has been the shaping of local human environments by political forces in ways that have allowed major economic actors to flourish on a worldwide stage. Working together, powerful political and economic agents have served as the 'architects' of economic globalisation, significantly mapping the development path for countries like Indonesia (Figure 1.6).

Globalisation and development

Globalisation is often seen as a powerful force that has brought modernising effects to nations, especially since the start of the 1980s. **Foreign direct investment** (FDI) in emerging economies is usually accompanied by infrastructure improvements (road, rail and telecoms). Initial injections of foreign capital often have virtuous knock-on effects for health and education in developing societies. In most parts of east and southeast Asia — the world region most frequently cited as the greatest beneficiary of globalisation — life expectancy has risen from 45 to 65 since the end of the Second World War. In contrast, poorer landlocked countries of central Africa — where far less foreign investment is received — still show low **human development index** (HDI) scores, meaning that there has been little real improvement over the last 60 years in life expectancy, literacy rates and purchasing power. Unsurprisingly, then, globalisation is often mentioned in the same breath as the phrase 'economic development'.

Figure 1.7 Rostow's model: the stages of economic development

Stage 5 High mass consumption
Consumer-oriented, durable goods flourish; service sector becomes dominant

Stage 4 Drive to maturity
Diversification, innovation, less reliance on imports, investment

Stage 3 Take-off
Industrialisation, growing investment, regional growth, political change

Stage 2 Transitional stage
Specialisation, surpluses, infrastructure

Stage 1 Traditional society
Subsistence agriculture, barter

According to Rostow, development requires substantial investment in capital. For the economies of developing countries to grow, the right conditions for such investment would have to be created. If aid is given or FDI occurs at Stage 3, the economy needs to have reached Stage 2. If Stage 2 has been reached, then injections of investment may lead to rapid growth.

Decades

W. Rostow was a prominent economist who devised a five-stage economic development model (Figure 1.7). Published in 1960, it pre-dates common usage of the word 'globalisation'. Yet the language used — notably, the phrases 'take-off' and 'mass consumption'— dovetails well with mainstream views about globalisation. For instance, when World Bank economists argue that FDI injections, or IMF and inter-governmental loans, are essential tools for poorer nations, there is a clear 'take-off' goal in mind. Integration into the highly interconnected capitalist world economy will, they believe, build prosperity and self-sustaining economic growth cycles capable of lifting a country out of poverty (Figure 1.8).

Figure 1.8 Self-sustaining growth cycles: the cumulative causation model of Gunnar Myrdal (1957)

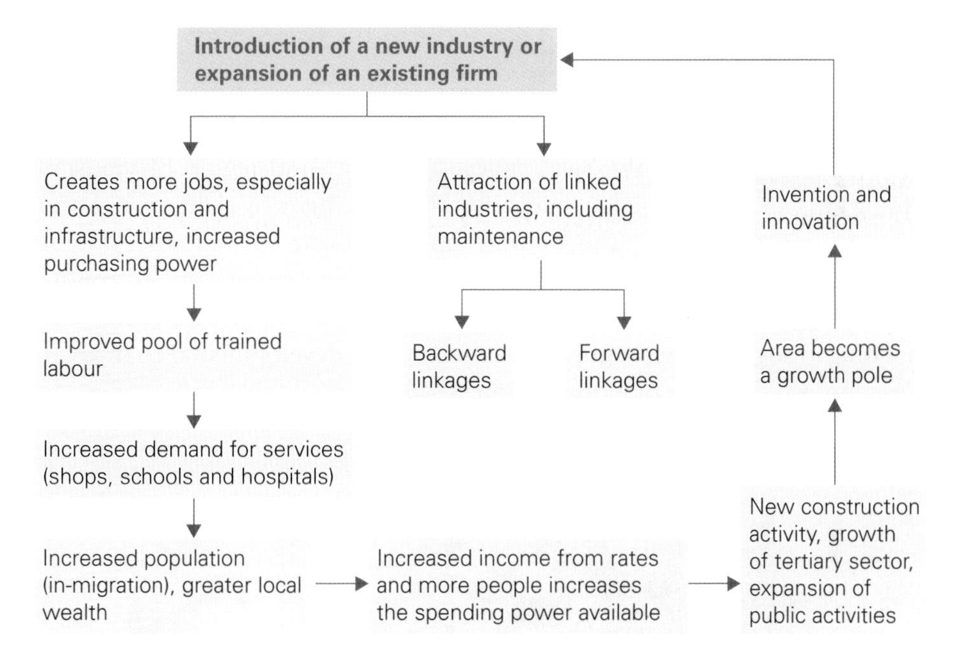

Writing in the late 1990s, Manuel Castells described globalisation as the development and growth of 'global networks' comprising interconnected and 'switched-on' places (see Chapter 3). Switched-on places are economically developed locations where fortuitous circumstances (such as an abundance of natural or human resources) invite investment flows which then transform these places into important global hubs for either the production or the consumption of goods and services, or for both. For instance, billions of dollars of overseas money have been invested in factories and workshops close to China's seaboard. With plentiful cheap labour and China's abundant supplies of coal on hand, this has become an optimum location for outsourcing and offshore investment by US, EU and Japanese firms. The changes under way have brought significant numbers of Chinese citizens greatly increased purchasing power. In 2008, UK-based department store chain Marks and Spencer opened its first store in Shanghai, signalling that the company now views China as a site for the consumption of its goods and not just for their production (even if, in what is still a 'two-speed' society, only a minority of Chinese people can afford to shop there). Many aspects of these economic changes are mirrored in India.

Any nation's long-term development has important non-economic aspects to consider too, such as social and educational improvements or enhanced human rights. These may, or may not, accompany the economic changes associated with globalisation. India now produces millions of university graduates each year, while in many parts of southeast Asia the gender gap has lessened, as more women enter waged work or even political life (Figure 1.9).

Yet China remains a communist party dictatorship, where internet use is heavily censored. In the Swat valley, a remote rural region of Pakistan, young girls are now denied an education that was theirs by right a generation ago. Around 200 girls' schools were burned down by Taliban militants during 2008–09. The violence is fuelled by a desire among some to counter 'globalising' influences perceived to be at work in the region. Quite clearly, it is wrong to suggest that progressive non-economic developmental changes are a guaranteed outcome of globalisation.

Figure 1.9 Skilled Chinese women workers assemble computer notebooks

www.flickr.com/photos/scobleizer/3009516045/

Indices of globalisation

Some attempts at describing globalisation are accompanied by actual empirical measurement. We will look at the KOF and then the Kearney rankings.

The Swiss Institute for Business Cycle Research, also known as KOF, is a respected part of the Swiss Federal Institute of Technology. For many years, it has produced an annual **KOF index of globalisation**. According to this index, in 2006 the USA and Sweden were the world's most globalised countries (Figure 1.10). A complex methodology informs each report (Table 1.1). Levels of economic globalisation are calculated by examining trade, FDI figures and any restrictions on international trade. Political globalisation is factored in, for instance by counting how many embassies are found in a country and the number of UN peace missions it has participated in. Finally, social globalisation — defined by KOF as 'the spread of ideas, information, images and people' — is accounted for, along with a cultural element; data sources include levels of internet use, television ownership and imports and exports of books.

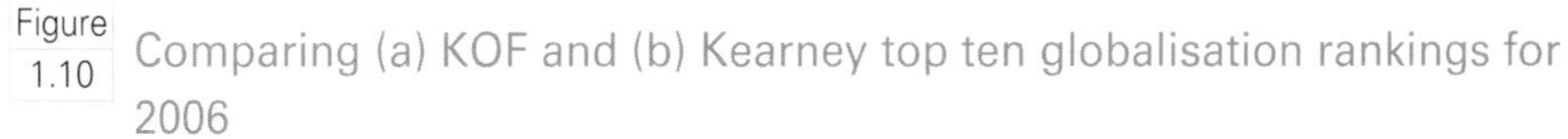

Figure 1.10 Comparing (a) KOF and (b) Kearney top ten globalisation rankings for 2006

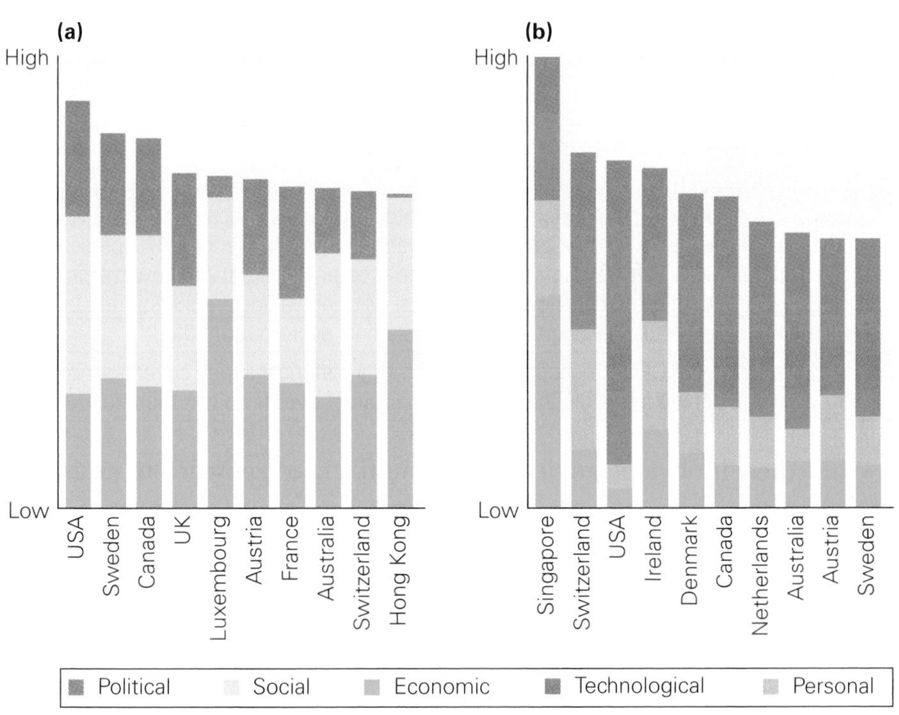

Table 1.1 Calculating the KOF index of globalisation

Stage	
1	To study economic globalisation, information is collected showing long-distance flows of goods, capital and services. Data on trade, FDI and portfolio investment are studied, including figures from the World Bank.
2	A second measure of economic globalisation is made by examining restrictions on trade and capital movement. Hidden import barriers, mean tariff rates and taxes on international trade are recorded. As the mean tariff rate increases, countries are assigned lower ratings.
3	Political globalisation data are looked at next. To proxy the degree of political globalisation, KOF records the number of embassies and high commissions in a country as well as the number of international organisations of which the country is a member and also the number of UN peace missions a country has participated in.
4	The assessment of social globalisation starts with use of personal-contacts data. KOF measures direct interaction among people living in different countries by recording (i) international telecom traffic (traffic in minutes per person), (ii) tourist numbers (the size of incoming and outgoing flows) and (iii) the number of international letters sent and received.
5	The study of social globalisation also requires use of information-flow data. World Bank statistics are employed to measure the potential flow of ideas and images. National numbers of internet users (per 1000 people) and the share of households with a television set are thought to show 'people's potential for receiving news from other countries — they thus contribute to the global spread of ideas'.
6	Finally, 'cultural proximity' data are collected. By KOF's own admission, this is the dimension of social globalisation that is most difficult to measure. The preferred data source is imported and exported books. 'Traded books,' explains KOF, 'proxy the extent to which beliefs and values move across national borders.'
7	Now all the data have been collected, a total of 24 variables (covering economic, political and social globalisation) are each converted into an index on a scale of one to 100, where 100 is the maximum value for each specific variable over the period 1970–present and one is the minimum value. High values denote greater globalisation. However, not all data are available for all countries and all years. Missing values are substituted by the latest data available. Averaging then produces a final score out of 100.
8	Each year's new KOF Index scores are added to a historical series covering more than 30 years, beginning in 1970. Changes in globalisation over time can then be studied.

While there is obvious merit in KOF's multi-strand approach to measuring globalisation, there are grounds for criticism too. For instance, does possession of a television set make a family more globalised, especially if its members only watch domestic output such as locally-made soap operas? Equally, the reasons why some countries volunteer large numbers of troops for UN missions are complex; it need not necessarily be the case that the most economically globalised countries are the most proactive.

Another well-known measure of globalisation is the **A. T. Kearney globalisation index**, an annual study assessing the extent to which around 70 of the world's most populated nations are globally connected (see Figure 1.10). Four strands of globalisation are investigated: economic, personal, technological and political integration. Data on trade and FDI inflows and outflows are used to calculate economic connectivity. Measurements of 'personal globalisation' are made with data showing levels of international travel and tourism, international telephone calls and cross-border remittances by migrants. Points for technological connectivity are scored according to the number of internet users and internet hosts found in a country. Finally, the political strand is evaluated by recording each nation's membership rating for a variety of international organisations, its financial contributions to UN peacekeeping missions and its ratifications of key multilateral treaties (such as the Kyoto Protocol).

When calculating the final scores, Kearney researchers add triple weighting to FDI and double weighting to trade volumes, due to those factors' particular importance in the 'ebb and flow of globalisation'. This is a value judgement made by A. T. Kearney, and other researchers might not want to treat the data in quite the same way. (It is worth noting that A. T. Kearney is an American private-sector management consulting firm, whose clients include many of the world's largest TNCs, such as ICI, BP Chemicals and Chevron Texaco.)

An additional weakness is that the statistics used by both Kearney and KOF may not always be reliable. Both surveys use a wide range of data, and methodological problems are certain to arise for many of the individual measures included. Crude averaging and statistical gaps further compromise the results. While they are an interesting starting point for the study of globalisation, both indexes lack the rigour and trustworthiness of, say, properly sampled and peer-reviewed scientific surveys of climatic data for different nations.

Activity 3

Calculate globalisation scores for people you know, either in college or at home. You can adopt the criteria used by either KOF or Kearney. Alternatively, you might devise your own multi-strand index. The easiest data for you to collect might be: the distance a person's clothes have travelled (the label usually indicates where a garment was made); the number of foreign destinations a person has visited; how many foreign-made television shows a person regularly views; or the overseas charities that an individual has contributed to.

Degrees of integration

We can identify a number of broad differences in national levels of global integration (Figure 1.11). First, there are those places that, for the most part, experience only one-way economic traffic. This is the shallowest form of integration, where local people have been corralled into cash-cropping or sweatshop activity, yet still lack sufficient wages to become significant consumers of global goods. TNCs see such places as producer and not consumer (market) nations. An example is Ethiopia, where coffee growers often live off less than one US dollar a day. Receipt of aid flows is another form of shallow integration. The poorest member states of the British Commonwealth are certain ex-colonies of the UK, such as Zambia, which are highly dependent on bilateral aid.

Figure 1.11 Differing levels of global integration

	Economic	Cultural	Political
Shallow	Simple one-way flows, e.g. receipts of aid or exports and sales of raw materials and cash crops	Western influences begin to overtake indigenous culture, changing diets, clothing and language	A country does not belong to influential global groupings (e.g. G20) and has limited voting powers
Deeper	Consumer-orientated TNCs open up stores in a country, as well as manufacturing products there; or a country has its own TNCs that begin investing in other places	Indigenous culture meets with imported fashion and food, mixing to create new hybrid forms; or a country's culture starts to exert its own global influence	Has G20 membership, or a seat on the UN Security Council, or other global political influence

'Two-way' economic connections develop when a nation's growing purchasing power begins to attract the attention of the giants of global consumerism — Coca-Cola, McDonald's, the cigarette manufacturers and even supermarkets and chain stores such as Wal-Mart. McDonald's has stores in 118 of the world's 193 countries: what does this tell us about market or other conditions in the remaining 75?

Once deeper economic integration begins, shallow effects of cultural globalisation can become more evident. Indigenous food and clothing may start to give way to more globalised (some would say 'Americanised') products. Diets in parts of Asia have become more meat and dairy based in recent decades, in part reflecting the powerful influence of firms like Burger

King. But this shallow cultural globalisation can be the precursor of deeper and richer forms of exchange. A nation's indigenous culture may begin to exert a reciprocal influence over foreign arrivals. A kind of conversation or dialogue may begin, giving rise to new hybridised or 'glocalised' cultural forms (see Chapter 8). For instance, Hollywood directors have for many years absorbed influences from Japanese, Chinese, South Korean and Indian film-making. Cuisine is another area where nations as diverse as Mexico, Jamaica, Greece, Italy, Japan, China, India and Thailand exert enormous global influence with their native food, a fact not altered by the arrival of American fast-food chains in these same countries.

Figure 1.12 Powerful global groupings

G20 'Group of Twenty'
(now 22 members, first formed in 1999)

G8 + 5 (first met in 2005)

G8 'Group of Eight'
(now 9 members, formed 1975)

France
West Germany
Italy
Japan
UK
USA
Canada (1976)
Russia (1997)
EU
Brazil
China
India
Mexico
South Africa
Argentina
Australia
Indonesia
Netherlands
Saudi Arabia
South Korea
Spain
Turkey

The interdependence of our destinies makes it necessary for us to approach common economic problems with a sense of common purpose *(1976 Joint Declaration)*

The information technology revolution and the globalization of markets have increased economic interdependence. *(1988 Economic Declaration)*

Finally, levels of political integration and global influence vary enormously. Few nations have a seat at the top table of international politics or are members of the important G8 or G20 groupings (Figure 1.12). Finance ministers of these countries jointly make key decisions affecting the entire global financial and banking system. Together, they hold close to 40% of the votes at the IMF and the World Bank and, from a diplomatic point of view, all have permanent influence on the United Nations' Security Council.

2 Technology and a shrinking world

Globalisation and communications technologies

Improvements in both the speed and capacity of transport and ICT (information and communications technology) are frequently cast as the key 'driver' in histories of globalisation. Important developments of the last 30 years — the internet, mobile phones, low-cost airlines, to name but a few — have certainly accelerated the process. Economic transactions are now easier to complete, be they productive (manufacturers can contract and outsource physical goods from increasingly faraway places, utilising ever-faster transport networks) or consumerist (goods, information and shares can be bought anywhere, anytime, online at the touch of a button). People move about more easily than in the past, too: budget airlines bring a 'pleasure periphery' of distant places within easy reach for the moneyed tourists of high-income nations.

Heightened connectivity changes our conception of time, distance and potential barriers to the migration of people, goods, money and information. This perceptual change has been described as time–space convergence (Janelle, 1968) and more recently as time–space compression (Harvey, 1990). Janelle plotted changing travel times between Edinburgh to London during Britain's Industrial Revolution and found that a two-week stagecoach journey in 1658 was ultimately superseded by an air flight lasting barely 2 hours. He concluded that different places 'approach each other in space–time': they begin to feel closer together, as each successively improved transport technology chips away more minutes and hours from the duration of the connecting journey. Since the sails of ships first filled with air, human society has experienced a 'shrinking world' (Figure 2.1).

In Harvey's account, time–space compression is crucial for the continued exercise of economic power. Technology has been pressed into service by global economic empire-builders. Fast trains and broadband connections are thus an outcome of the never-ending search for new markets and profits by

Figure 2.1 A shrinking world: the time taken to circumnavigate the globe

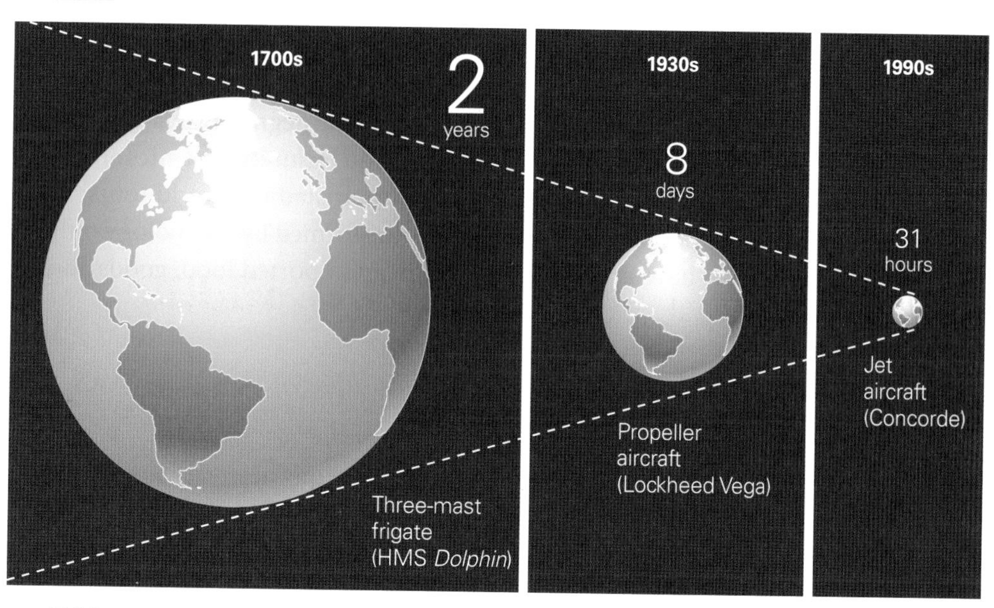

TNCs. A second axis of human power — the exercise of military might and imperial ambition — also continues to provide an equally important stimulus for transport and communications innovation. One of the earliest shrinking-world technologies served nations' security: the practice of lighting warning fires across a chain of beacon hills dates back to ancient Greece. As recently as 1588, the first sighting of the Spanish Armada was signalled across England in hours by hilltop fires.

This military imperative to develop new technology is also evidenced by twentieth-century design. Jet engine science was refined during the Second World War and the 1950s Korean conflict. Communication satellites and major advances in GIS/GPS owe much to the Cold War (after the Soviet Union launched Sputnik in 1957 and entered into a 'space race' with its rival superpower, the USA). The origin of computing also lies in wartime research and development, including the British Colossus (1943) and German Z-3 (1941) projects, and the internet itself has important military roots.

Activity 1

Research the most recent historical events that were marked in the UK with beacon fires. Using OS maps, investigate whether there was any such beacon near your school or college. (Note how the word 'beacon' survives in relief features sharing the common name 'Beacon Hill'.)

Power and place

Doreen Massey is a contemporary geographer who has written critically about changing perceptions of place in a technologically advancing world. She has argued that time–space compression is socially differentiated: not everyone experiences the sense of a shrinking world to the same degree. Privileged elite groups fly around the world for reasons of work and leisure: academics attend international conferences and rock stars enjoy their stadium world tours. In contrast, many more people's lives have been transformed by technology only in so far as they now have access to a glut of cheap imported food, goods and television shows.

The technology timeline

Historical studies of transport and communication innovation provide milestones by the dozen (Figure 2.2). Four especially significant postwar innovations that have increased interactions between places are listed opposite.

Figure 2.2 A transport and communications timeline

3000 BP The Egyptians have ships under sail as early as 3200 BP. Ancient Greeks light beacon bonfires on hilltops; the Romans later establish chains of high posts with fire-baskets that allow basic alarm signals to be sent across country.

1500s–1700s Industrial canals and stagecoach routes begin to reduce land travel times for goods and people (without each individual needing a separate horse).

Early 1800s First steam ship crosses the Atlantic in 1819 (29 days). The telegraph is the first technology to instantly send short messages by Morse code across great distances, even between continents after 1866.

Late 1800s The first telephone call is made in 1876, followed in 1896 by voice transmission using radio.

Early 1900s The Ford motorcar is launched in 1903. The first television is built in 1926 while Boeing begins commercial flights in 1928. During the 1940s, important steps forward are made in computer technology and also in the manufacture of transistors.

1970–1980s Mass-produced desktop PCs start to appear, with laptops following soon after. Mobile phones arrive in 1973 but are extremely large. From 1989, the World Wide Web (internet) is developed.

1990–2000+ The internet takes off after search engines appear (Google is founded in 1996). People quickly begin communicating via modem and later broadband. Mobiles become smaller and cheaper. i-Pods, blogs, MySpace, Facebook and Twitter all capture the public imagination.

- **Container shipping** Some commentators describe this, rather than the internet, as the real backbone of the global economy. With 200 million shipping container movements made every year, shifting everything from chicken drumsticks to patio heaters across the planet, this technology has, since its take-off in the 1950s, helped TNCs bolt together expansive production and consumption networks.
- **Lorries** Manufactured goods are the lifeblood of consumer societies. In the USA, coursing through its interstate veins, 25-metre long megatrucks keep the retail parks fully stocked with goods 365 days of the year (see Figure 2.3). In the UK, supermarket giant Tesco has plans for an 85 000 square metre 'MegaShed' depot close to the M3 motorway that will serve as a hub for its own dense network of lorry movements, on the move every hour of every day.
- **Air travel** Like the USA and the EU before it, India is currently experiencing a transformation in terms of its internal connectivity. The Indian government recently ring-fenced US$12 billion for the building of new airports. Estimated rising demand for internal air flights from India's new middle classes will require the number of planes in service to rise from 200 in 1991 to 2000 by 2020.
- **High-speed rail** Railways are the chief conduit linking rural and urban parts of China. Migrant workers travel in both directions along the route of the 1500-kilometre China–Tibet 'sky train', whose hi-tech specifications mean it can operate effectively even in the conditions on the Tibetan plateau where temperatures drop to –35 °C.

Figure 2.3 Megatruck on an interstate in the USA

www.flickr.com/photos/tomsaint/2891788700/sizes/l/

Evolving alongside improved means of transport for people and goods are information storage, retrieval and transfer technologies. ICT enables global movements of corporate and personal data, while increasingly serving as the medium of choice for people to consume music and film (in preference to purchasing manufactured CDs and DVDs that require physical shipping to high-street stores before they can be sold). There have been many important breakthroughs in the science of data processing and diffusion over the last 50 years. These are typically rooted in just a handful of key geographical locations, where both formal and informal quaternary research clusters are found:

- **The transistor** Invented at Bell Laboratories in New Jersey in 1947, with the later shift to silicon accomplished by Texas Instruments in Dallas in 1954, the transistor is regarded by some historians as the most important invention of the twentieth century — and, by implication, for globalisation. Transistors are used to switch electrical current on and off, allowing computers to process binary data. Leading Bell transistor scientists later helped create the Silicon Valley research cluster in northern California during the 1950s.
- **The microprocessor** Introduced by Silicon Valley's Intel Corporation in 1971, microprocessors today are small devices inside computers that contain as many as two billion miniaturised transistors.
- **The personal computer** The first small-scale computer designed around a microprocessor was named Altair. Built by a small Albuquerque calculator manufacturer in 1975, Altair served as the basis for Apple microcomputers (designed by Steve Wozniak and Steve Jobs in Silicon Valley). Archrival IBM introduced its own PC (personal computer) in 1981.
- **Windows** Icon-based user-friendly interface technology and software was introduced first to the Apple Macintosh in 1984 and to the PC by Microsoft as Windows 1.0 in 1985. Microsoft was founded in Albuquerque and later moved north to Washington state.

Mobile phone uptake in Africa

The mobile phone is worthy of attention, given the leading role it has played in 'switching on' the planet, including many of the world's poorest societies. In 2009 there were an estimated 4.3 billion mobile subscriptions worldwide: roughly six in ten people alive today. Mobiles are increasingly affordable to all but the world's very poorest subsistence and slum communities. Market penetration across the African continent soared from just 2% in 2000 to 28% by 2007 (Figures 2.4 and 2.5).

Figure 2.4 Global variations in mobile phone uptake over time (mobile phone subscriptions per 100 inhabitants, 1997–2007)

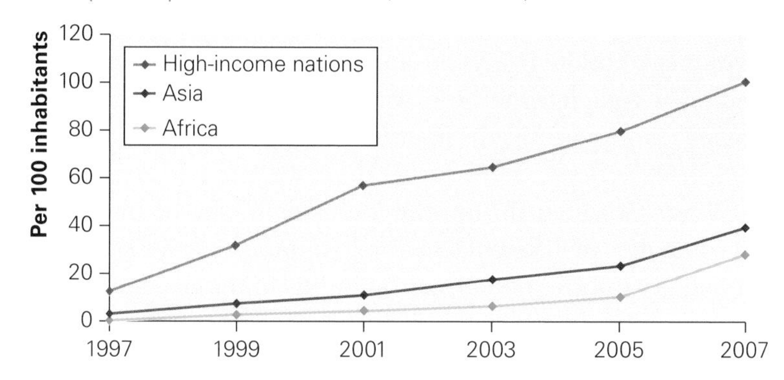

In countries where the lack of any effective communications infrastructure has traditionally been one of the biggest obstacles to economic growth, cellular technology is changing lives:

- Money can now be directly transferred between phone users (5 million Kenyans use Vodafone's M-Pesa money transfer service).
- Fishermen and farmers use mobiles to check market prices before selling produce.
- Greater mobile uptake can support democracy: political parties have a means of getting their message across to more people.

Other notable technological changes across Africa include the recent arrival of innovative wind-up radios designed by Trevor Baylis. Cheap laptops for children are championed by many organisations, including OLPC (One Laptop Per Child) and Intel. Although access to the internet generally remains poor in east Africa, Kenya's growing call centre industry is starting to attract interest and investment from the fibre optic cable companies that build global broadband networks.

Figure 2.5 Monitoring share prices on the Kenyan stock exchange using a mobile

www.flickr.com/photos/whiteafrican/2594981758/in/set-72157605582824438/

Evolving communications networks

Transport and communications network growth has often been driven by state investment, both in peacetime and in time of war. The first trans-Atlantic telegraph cable, completed in 1866, was subsidised by both the British and US governments. More recently, the internet began life as part of a scheme funded by the US Defense Department during the Cold War. Early computer network ARPANET was designed during the 1960s as a way of linking important research computers in different locations. (Rumours persist that it would have allowed military information to survive online in the event of the destruction of key command and control centres by hostile Soviet forces.)

Today, powerful TNCs continue the process of network-building first begun by states. For instance, Google recently commissioned its own multi-terabit trans-Pacific fibre optic cable. The rapid development of commercial broadband networks can therefore be described as the result of a combination of public and private investment. It is still a work-in-progress, for both poorer countries and richer ones, with poverty precluding connectivity in many places (Figure 2.6).

Figure 2.6 The uneven global distribution of undersea cables and internet access

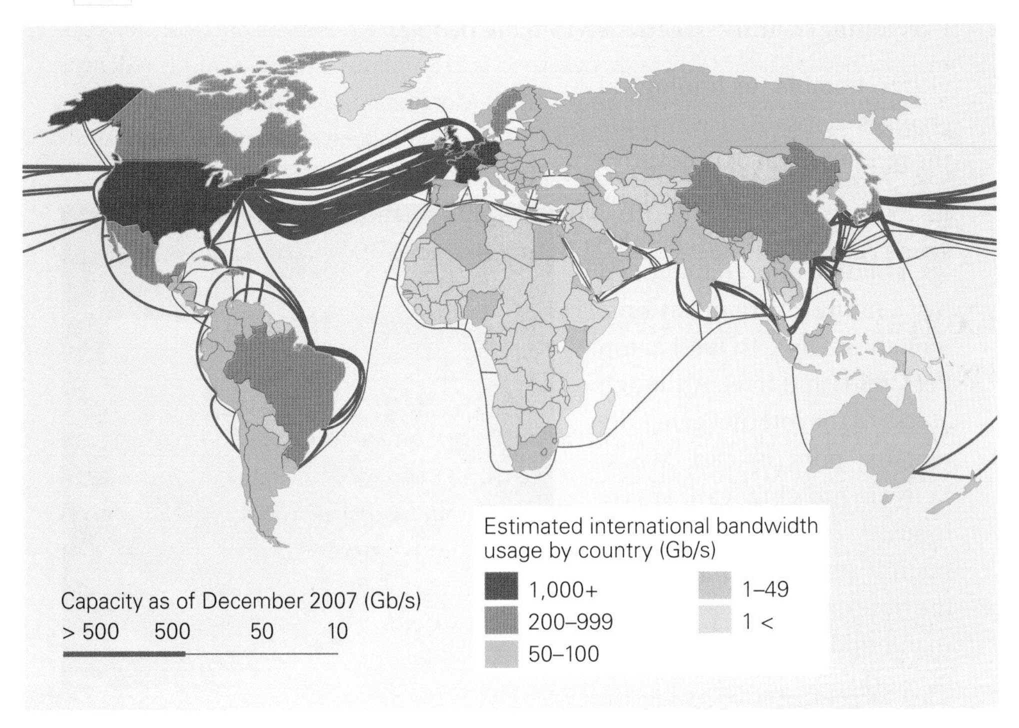

Activity 2

Study Figure 2.6.

(a) Estimate the capacity that exists between: Europe and the USA; South Africa and Asia; Japan and North America.

(b) Describe the pattern of total broadband usage shown. Suggest reasons for the regional contrasts that emerge.

However, wherever internet access is available and allowed, and where incomes support subscription fees, community membership has ballooned since the 1990s. Online social networks have flourished off the back of this, although their durability is still hard to judge. They can be entirely ephemeral: short-lived Facebook communities spring into existence simply to help people coordinate their attendance at music festivals or sporting events. The 2009 G8 summit held in London was disrupted by anti-globalisation demonstrators who used social networking sites Facebook and Twitter to organise themselves. Deployment of new social networking technologies to create 'flashmobs', and other expressions of local, anti-global radicalism, is a theme for Chapter 8.

Along the way, telecoms network builders have overcome enormous challenges set by physical geography. They have slung mile upon mile of fibre optic cable across the abyssal plains of the ocean floor. The network mostly holds firm, although beset by currents and the occasional volcanic eruption. Capricious nature may still occasionally pull the plug. In 2006, a submarine earthquake and landslide destroyed Taiwan's telecom link with the Philippines. Human error can also creep in; Asia temporarily lost 75% of its internet capacity in 2008 when a ship's anchor severed a major internet artery running along the seabed from Palermo in Italy to Alexandria in Egypt.

Communications liberalisation and network expansion

The invention of faster forms of transport and communications does not necessarily guarantee increased ease and speed of international movement for people, goods or data. For instance, state censorship in China restricts people's day-to-day use of the internet; Chinese citizens cannot view CNN or BBC news websites. Yet this is an unusual case. For the most part, data now flow freely between nations where technical capacity exists for internet access, rendering financial transactions virtually effortless — be it book-buying at www.amazon.com or online share-dealing. The physical movement of goods, both raw materials and manufactured consumer products, has also become much easier in recent years thanks to the liberalisation of trading laws. Multi-governmental trading blocs such as the EU and NAFTA (see Chapter 7) have taken shape alongside the creation of local export areas called **special economic zones** (SEZs). Both encourage transnational movement of goods by slashing import and export duties, tariffs and quotas. All of this has provided an impetus for more large firms to move their production wings offshore and overseas into new least-cost locations.

In Europe, nations have legislated for international journeys of people and goods to take place effortlessly. Unprecedented liberty of movement is found throughout the region, allowing transport providers such as easyJet and Eurostar to flourish. EasyJet's market has grown from 30 000 passengers in 1995 to 44 million in 2008 across 387 European and North African routes. Eurostar runs its services through the Channel Tunnel linking England and France, which opened in 1994. Both firms carry people across Europe without the inconvenience of customs, making a continent feel more like a country. Of course, the free movement of people granted by the EU's 'four freedoms' is an exceptional case. In North America, manufactured goods crossing the Mexican–US border are not accompanied by people, save for the truck drivers. Barbed wire fences and border police patrolling the American side in search of illegal migrants testify that this is not yet an entirely borderless world.

Technology and the offshoring and outsourcing of services

Offshoring describes the relocation of a firm's production to another country, or new investment abroad rather than in the home country. The work may be outsourced to a foreign company or else a new wholly-owned branch plant or office may be built. Chapter 3 examines this trend in more detail but some mention must be made here of the crucial role technology has played in the process. In particular, broadband internet has allowed a range of new worldwide endeavours — such as international call centres and global media collectives — to flourish. Now that enormous volumes of data can flow in real time via the fibre optic cables that link continents, all manner of new shrinking-world enterprises are emerging:

- BBC Worldwide launched the animated series *Freefonix* in 2008. The scripts were written in the USA and voices recorded in the UK, before a Paris team modelled the characters. Next, the soundtrack and outline

sketches were sent to Trivandrum in India for animation prior to post-production editing in Ireland. Large bundled media files shuttle between these different **creative hubs** via fibre optics.
- 'Gold farmers' is the name given to Chinese employees of Wow7gold, a company that helps Western teenagers succeed at playing the computer game *World of Warcraft*. Workers earn around £100 a month completing hard parts of the game for paying customers. These 'playbourers' face long shifts, often working through the night.
- Another interesting case study is the offshoring of examination marking. Increasingly, schools examination boards and agencies throughout the world are using technology to increase efficiency, cut costs and, they hope, improve the accuracy of marking. Intelligent scanning of scripts allows images of answers for each separate exam question to be sent via broadband internet to different locations around the world for specialist or non-specialist online marking.

Activity 3

Many major action feature films are the result of international collaboration, with different studios across the world contributing to the finished production.

(a) Study the closing credits of special effects 'blockbusters' for evidence of this.

(b) Using an outline world map, plot the locations of the studios you have identified.

Technology and society

Is the role that technology plays in the story of globalisation best understood as a *cause* of major worldwide social and economic changes? Great leaps forward in the history of communications technology can be explained in similar terms.

Many important discoveries have actually been need-driven and are thus an *outcome* of people's efforts to solve existing problems (Figure 2.7). Thus many would argue that it is the economic needs of TNCs that have driven technological innovation in the era of globalisation. The crisis of global capitalism experienced during the 1970s — when rising production costs for goods coincided with a period of falling wages and declining consumption in the world's economic cores — gave an impetus for many firms to invest in the research and development of new products. They also actively sought new and improved ways of moving parts, goods, money and people around the world

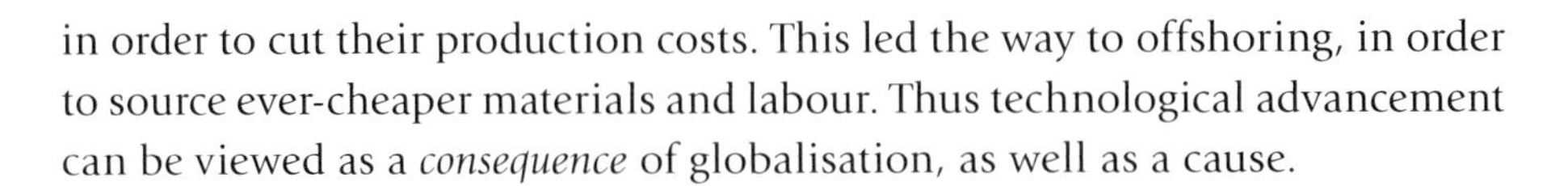

in order to cut their production costs. This led the way to offshoring, in order to source ever-cheaper materials and labour. Thus technological advancement can be viewed as a *consequence* of globalisation, as well as a cause.

Figure 2.7 Two views on why technology changes occur

Technology
Communication and transport technologies have been improving for thousands of years; many breakthroughs have been made by inventors who are interested in science simply for its own sake — not just to serve society

Social need is the 'hard force' that drives technological change — but only when specific human needs arise that are not met by existing technologies

Technological progress is the 'hard force' and it occurs for independent reasons — new inventions then trigger social and economic change in turn

Society
Capitalist economies always seek to increase profits, by using transport and communications to build new global markets — military needs constantly drive innovation; so do social needs, including maintaining friendships and education

Following this logic further, we understand that communications network-building is a process that will never be fully completed. Even if poverty and other obstacles to technological access around the world could be wished away, market-driven technical innovation must continue — in order to create new product lines or to design faster ways of moving goods and selling them online. Speedier broadband, superior handheld wireless devices, higher-definition television screens: we may expect perpetual design enhancement, far into the future. Each innovation, as it arrives, will be rolled out first where markets are strongest: in affluent global hubs. Peripheries will once again find themselves lacking access due to weaker demand, thereby reawakening political concern over 'two-speed' access to technology, or a 'digital divide'. It is a challenge that will need to be tackled time and time again.

3 Transnational corporations and global economic networks

Financial flows are of primary importance in any account of globalisation. Movements of money are channelled around the world by many different players, including TNCs, nation states, multilateral agencies such as the World Bank, non-governmental organisations and charities. Of these, TNCs are the key agents — or 'architects' — of globalisation in the twenty-first century, transferring enormous amounts of capital between places in the form of foreign direct investment (FDI). The great global increase in interconnectivity and interdependency witnessed over the last 30 years owes more to FDI than to any other type of monetary flow, with offshore investment by TNCs rising near-exponentially from US$20 billion in 1980 to US$1700 billion in 2008.

The diversity, scale and role of TNCs

A transnational corporation is a firm that has coordinated operations based in more than one country. By most criteria, TNCs are a diverse category:

- **Market sector** Some TNCs work in primary industries consisting of food and forestry (the agribusiness sector), oil and mineral extraction. Others, including about a hundred of the world's most powerful TNCs, such as Toyota and Citigroup, dominate global manufacturing and banking.
- **Scale and size** There are around 60 000 TNCs (defined as companies with operations in more than one country). The top 100 own 20% of world financial assets and enjoy 30% of global consumer sales.
- **History** Some firms have grown over time through mergers and acquisitions, in effect 'colonising' the firms of other nations, aided in this process by financial deregulation and trade bloc growth (Chapter 7).

Large firms also differ greatly in terms of their structure and function. Some retain highly centralised command and control functions; thus the provenance of all worldwide decision-making is located with a small but powerful group of people stationed in corporate global headquarters. Other firms have regionalised their decision-making, with relatively autonomous local branches or subsidiary companies. Good examples of this latter approach are Unilever's Indian operations and Procter and Gamble in China.

Accountability for overseas operations, and their social and environmental impacts, varies according to the precise nature of the **spatial division of labour** a firm adopts. Some TNCs own their overseas operations and are held directly responsible for the treatment of labour on the shop floor. For instance, in 1987, American guitar manufacturer Fender established a major branch plant in Mexico. This facility — one of Mexico's *maquiladoras*, or export assembly plants — is directly owned by Fender. Its employees receive their wages directly from the company and the famous Fender logo is prominently displayed above the factory's entrance.

In complete contrast, many other firms contract out much of their work to third parties. In 2008, American clothing giant Primark found its own ethical code of conduct ('anti-sweatshop' guidelines for the treatment of workers) was being ignored by three of its major Indian subcontracted manufacturers. More complex yet, a subcontractor employed by a TNC will sometimes act as a middleman, outsourcing the work in turn to an even cheaper and less well-regulated manufacturer. This further lengthens the supply chain, muddying the waters of social responsibility in the process.

Such **shadow factories** — operating undetected but in close proximity to approved suppliers — reveal much about the convoluted geography of TNCs. In China, Indonesia and elsewhere, use of shadow factories is often routine. Subcontracting information is not always made available to consumers, and ethically conscious shoppers can be left unaware that they are purchasing goods manufactured by children or poorly treated adult workers.

The largest TNCs possess great wealth held in shares, stock, assets and profits. In 2008, Wal-Mart had revenues of US$379 billion — more than six times the gross domestic product (GDP) of Bangladesh, a nation of 150 million people. But despite being enormously rich, TNCs are not all-powerful.

- In 2009, the Chinese government blocked Coca-Cola's proposed takeover of China's largest fruit drink manufacturer, Huiyuan Juice.
- Venezuela's leader Hugo Chávez has insisted that foreign oil companies transfer ownership of their Venezuelan operations to the national government. In 2009, troops were mobilised to seize the assets of 60 TNCs, as Chávez set about severing ties with these firms and the IMF.

Activity 1

Table 3.1 The world's largest TNCs and the locations of their headquarters

Rank	Company	2008 Revenues (US$ billions)	Country of headquarters
1	Wal-Mart Stores	379	USA
2	Exxon Mobil	373	USA
3	Royal Dutch Shell	356	Netherlands/UK
4	BP	291	UK
5	Toyota	230	Japan

Study Table 3.1.

(a) Estimate the combined revenue for the firms listed and divide this total by the number of people living today (approximately 7 billion). What does the resulting figure tell you about the size, scale, influence and profitability of these TNCs?

(b) Research the figures showing the gross domestic income (GDI) for the world's five poorest nations. Present your results and the data given in Table 3.1 in graphical form so that comparisons can be made.

Impacts and accountability

There is controversy over the role TNCs play in promoting globalisation and the impacts they bring to poor people and places (Table 3.2). When things go badly wrong — as in the case of the 1984 accident at Union Carbide's poorly maintained pesticide factory in Bhopal that resulted in around 10 000 deaths — where does accountability rest? Union Carbide has since been bought by Dow, an even larger American chemicals firm. Dow has ignored repeated requests for financial compensation made by Bhopal victims on the grounds that it cannot be held responsible for an accident in India occurring prior to its acquisition of Union Carbide.

To what extent does a TNC 'belong' to a particular place? In addition to the location or relocation overseas of large firms' manufacturing operations, their research, marketing and administration offices have also been scattered to the four winds (for instance, many UK banks now have call centres in India). Over time, the world's largest corporations have undergone a shift in ownership, so that that too has become transnational. Most are now typically owned by a blanket group of global shareholders investing through different regional stock markets.

Table 3.2 Opposing views about TNCs

The case against TNCs	The benefits TNCs can bring
■ *Tax avoidance.* TNCs may avoid paying full taxes in the countries where they operate, through transfer pricing and tax concessions. This means that governments find it harder to raise revenues, provide services and respond to the demands of local people. ■ *Limited linkages.* FDI does not always help developing world economies. If links are made with local firms (e.g. in the sourcing of raw materials) then more wealth may be generated. ■ *Growing global wealth divide.* By selectively investing in certain regions (e.g. southeast Asia) while largely bypassing others (e.g. sub-Saharan Africa), TNCs are active agents in creating a new geography of 'haves' and 'have-nots'. ■ *Environmental degradation.* TNCs are often a major cause of environmental degradation, which has the greatest impact on the poor. One of the most notorious cases of this occurred on 2 December 1984, when poisonous methyl isocynate gas was emitted from the pesticide plant in Bhopal, India, owned by the US TNC Union Carbide. This led to the deaths of thousands of Indian people living close to the plant.	■ *Raising living standards.* TNCs invest in the economies of developing countries. They are sometimes active in raising wages and can help spread wealth globally. FDI has helped put China on its way to becoming the world's second largest economy, overtaking Germany and soon Japan. (China is the world's largest recipient of FDI, with around half a million foreign-funded enterprises.) ■ *Transfer of technology.* TNCs can be responsible for the transfer of technology and managerial know-how. For instance, South Korean firms such as Daewoo and Samsung have learned to design, make and sell their own products to foreign markets. ■ *Political stability.* In eastern Europe and China, investment by TNCs has contributed to economic growth and political stability. This may be contrasted with conditions in much of Africa, where instability, civil war and distance from markets have made the investment environment less favourable. ■ *Raising environmental awareness.* Because large TNCs have a corporate image to uphold, they sometimes do respond to criticism. Many large firms are now trying to establish their 'green credentials' by starting to address issues around packaging, transport and carbon emissions, while also increasing their fair trade commitment.

Figure 3.1 Investigating to which country a TNC belongs

?

In which nation is the bulk of assets and senior staff located?

What is the nationality of the board of directors and decision-makers?

What is the legal nationality of the parent company?

To which nation would the group turn for diplomatic protection and support?

In which nation is the TNC as a whole taxed on its worldwide earnings?

Yet despite their seemingly 'global' identity, large TNCs can still be identified as economic agents of particular nation states if the right questions are asked (Figure 3.1). In some exceptional cases, TNC control — and accountability — is found to be invested in the hands of a few easily identifiable individuals. Wal-Mart famously remains under the influence of the Walton family, whose combined inherited wealth is estimated to be around US$80 billion. The Waltons are both majority shareholders and company directors.

The changing geography of TNCs

The geography of production

In the early years of globalisation, investments made by large firms overseas were simple 'clone' operations using locally found materials. From the 1920s onwards, a firm like car manufacturer Ford would seek to more or less replicate its entire US operation inside a subsidiary market, such as the UK. But after 1945, new technical developments, and the onset of a period of unprecedented geopolitical stability for developed nations, led firms to devise more complex cross-border systems. As a result, flows of parts for overseas factory assembly lines are now likely to originate from a plethora of international least-cost locations.

With so many multiple upstream and downstream linkages in the supply chain, it becomes impossible to pin down where a car or television set is really 'made'. Although the final point of assembly is recorded, an electrical consumer item may contain parts made in dozens of different countries. The Mini car manufactured by BMW in Oxford uses 1600 component parts sourced from all over the world. This is a far cry from the days of European car production during the 1960s when vehicles were mostly manufactured using locally sourced parts, from windscreen wipers to hubcaps.

The geography of consumption

Luxury manufactured goods are consumed in many more places today than in the past. Ben Thanh market in Vietnam's Ho Chi Minh City is a new consumer hotspot, where the cost of prime sites for retail outstrips central business district (CBD) land prices in Tokyo. UK supermarket Tesco has opened 300 stores in Poland since European Union (EU) enlargement in 2004, and is targeting South Korea and China as two highly promising emerging markets. Similarly, Sweden's Ikea has expanded its network of retail outlets over time, with 165 megastores now operating across 33 countries, while sourcing parts from 1600 suppliers in 55 countries. Owner Ingvar Kamprad was the world's fifth wealthiest individual in 2009.

The growth of BBC Worldwide (a global commercial wing of the UK television corporation) is another interesting case study of consumption. With production outlets in India and Los Angeles, BBC Worldwide has established itself as a major global TNC and is busily engaged exporting its flagship programming worldwide. The successful BBC show *Strictly Come Dancing* (re-branded as *Dancing with the Stars* in the USA) has been re-imagined and re-filmed in 40 territories.

Activity 2

Which television shows that you watch, or have heard of, are re-filmed for audiences in different countries? Conduct an online investigation to find out how many different territories are served by their own version of the shows you have identified. Also identify the media TNCs that own and license these programmes.

Distribution patterns

The distribution of TNCs around the world has changed beyond recognition since the 1990s, in large part due to the economic rise of India and China. There have been two main phases to these countries' growth. First, in the early 1990s, both nations became popular destinations for offshoring and outsourcing by foreign TNCs (China received US$45 billion of FDI in 2002). Second, China's and India's own companies have more recently emerged as major players investing in other places. Enormous outbound foreign investments are made each year by India's CFS and Infosys, and by China's Haier and Huawei. Landmark developments include:

- The 2006 buyout of Corus, previously British Steel, by Indian firm Tata — a move some commentators branded 'reverse colonialism'.
- China becoming a major investor in Africa, with Chinese FDI there totalling US$30 billion in 2007. In 2008, there were 750 000 Chinese people working in Africa for 900 Chinese companies.
- Indian TNC owners, such as Tata chief Lakshmi Mittal and petrochemicals tycoon Mukesh Ambani, now featuring in the 'top ten' world billionaire 'rich list'.

Global networks and flows

A network is an illustration or model that shows how different places are linked together by connections or flows, such as FDI made by TNCs. Network mapping differs from topographical mapping by not representing real distances or scale but instead focusing on the varied level of interconnectivity for different places, or nodes, positioned on the network map (Figure 3.2). Especially well-connected places — including world cities such as São Paulo, or research clusters in Stanford (USA) and Cambridge (UK) — are described as **global hubs** in network theory. Key economic network flows comprise the following:

- **Money** Major flows include TNC investment, international aid and loans to countries channelled through the IMF and World Bank. There are TNC

Figure 3.2 Places in (a) topographical and (b) network mapping

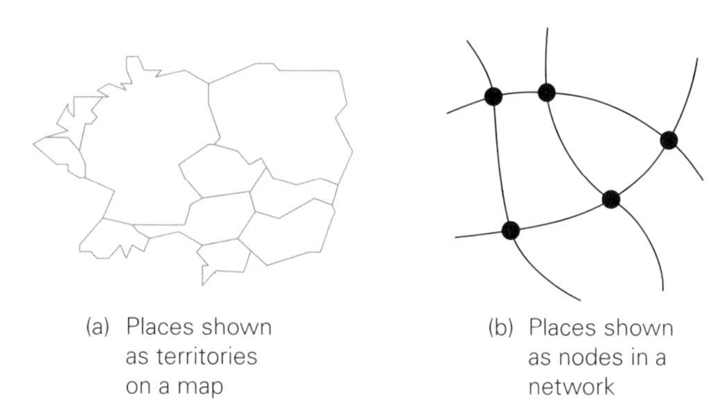

profit flows to follow too, the tracking of which is not always easy. Some firms have elaborate corporate structures that allow them to move money invisibly through subsidiary operations into offshore tax havens in places like the Cayman Islands. By such means, they minimise taxes paid in the countries where products are produced or sold.

- **Raw materials** Vital commodity flows include food, minerals and oil, channelled via the operations of major TNCs such as Cargill, Rio Tinto and BP. Frequent fluctuations in the cost of basic resources have profound impacts on quality of life for primary sector workers.
- **Goods and services** Around 200 million container movements take place each year on seafaring vessels. China — the source of so many manufactured goods flows — is often called the 'workshop of the world'.

Alongside these economic flows, two further important types of global movement are taking place, for reasons other than trade:

- **Information** One recent estimate suggested around 60 billion non-spam e-mails were sent daily by the 1.6 billion people online in 2008.
- **People** Each year around 200 million people are believed to migrate internationally, up from 80 million in 1965. This figure includes legal and illegal economic migrants (around 130 and 40 million respectively), as well as a smaller number of displaced persons (refugees escaping from physical disasters or political persecution).

Not all flows of goods, services and people between places are legal (Figure 3.3). Vast economic flows of illegal drugs circulate worldwide in all directions, while unregistered immigration, slavery and cross-border people-trafficking flourish despite the best efforts of authorities to prevent them.

Some kinds of information flow — including certain types of internet pornography or websites inciting racial hatred and violence — are deemed illegal in many countries, and in China online censorship is routine. However, effective internet policing is difficult to achieve in practice and causes enormous headaches for law enforcement agencies.

Figure 3.3 Prohibited flows and illegal flows

Prohibited flows

Cuba → USA

The USA imposed a **complete trade embargo** on communist Cuba in 1962 as a result of Cold War antagonism between the two countries. The result? A commercial and financial blockade.

World → China

Not all **information flows** are allowed to enter China. For instance, internet users there are not allowed access to the BBC's Chinese-language website service.

Australia → New Zealand

For 50 years, imports of Australian **honey** were banned in New Zealand for fears of a 'bio-security threat' (Australia's bees suffer from a disease that New Zealand beekeepers have been keen to avoid).

China → Europe

In 2005, the EU briefly banned the further imports of **cheap Chinese textiles** — especially women's bras — in an attempt to protect its own manufacturers. This was dubbed 'bra wars' by the media.

Illegal and criminal flows

Afghanistan → UK

By some estimates, 60% of Afghanistan's GDP may come from illegal **opium trade**, feeding the demand for heroin among drug-users on the streets of European cities.

Colombia → USA

99% of **cocaine** reaching the US is from Colombia, amounting to a billion-dollar trade. To fight this, the USA has in return given US$3 billion in mainly military aid to Colombia to fight the drugs trade.

Nepal → India

Girls as young as 10 years of age are kidnapped and taken to India where they are sold by **people traffickers** to brothels. There, the girls will work as prostitutes in a form of modern slavery.

Myanmar (Burma) → Thailand

Each year, 100 000 **illegal migrants** escaping repression and poverty in Myanmar are intercepted by Thai border guards and promptly returned to Myanmar.

Network flows, whether legal or illegal, have stimulated the imagination of writers and artists, including Chris Gray, who has represented the world in the style of the iconic London Underground network map. In Gray's world-view, the international borders that separate cities are no longer present, while the distances shown between places bear no relation to real-world kilometres: physical separation poses no obstacle to information flows between places in the internet age. The result is an isotropic surface of nodes and hubs in a borderless world, connected by multicoloured flow lines (Figure 3.4).

Figure 3.4 Europe redrawn as a network (based on the design of the London Underground)

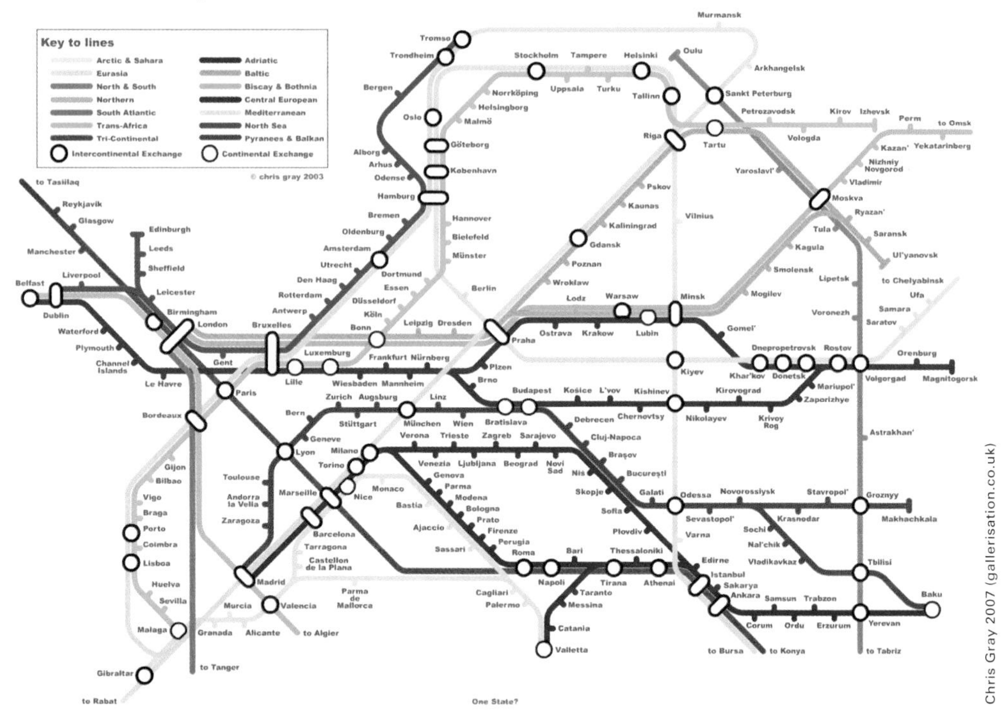

Chris Gray 2007 (gallerisation.co.uk)

Building interconnections

Network flows do not provide connectivity for all the world's places and people to anything like the same degree. Writing in 1996, globalisation author Manuel Castells described how the global economy was beginning to function in real time as a single unit, managed by key players, such as the major TNCs, which constantly scan the world for potential profit, connecting together skilled or low-cost populations and affluent markets. At the same time, unskilled labour markets, from Burkina Faso to the flooded ghettoes of New Orleans, are left 'switched off' from economic globalisation. A 'Fourth World' of switched-off places emerges from this network analysis: a world of social and economic exclusion, whose inhabitants — be they subsistence farmers in Chad or homeless Londoners — can be found in any continent or city.

This global map of economic inclusion and exclusion (and of interconnections, and lack of them) is more than a mere outcome of TNC actions and market forces. As we saw in Chapter 1, Indonesia's great leap forward after the 1960s was partly a result of US political support for General Suharto (Figure 1.6, page 13). This was not an isolated incident. Agencies such as the American CIA have supported regime changes that include the ascent of General Augusto Pinochet in Chile and Ferdinand Marcos in the Philippines. William Blum, a former US State Department official and the author of *Rogue State: a Guide to the World's Only Superpower*, has counted US involvement in at least 40 attempts to overthrow foreign sovereign governments since 1945. Notable US interventions include Colombia (1999), Nicaragua (1979), Angola (1975), South Vietnam (1963), and Cuba (1961). There has always been — and will continue to be — a major geopolitical dimension to global network-building.

Interconnectedness within economic networks in turn brings the benefits and costs of **interdependency** (the notion that what happens in one place has effects for other places). Becoming 'switched on' to global networks has brought economic prosperity to many, though by no means all, people. However, a heightened degree of risk is also introduced to well-connected places. This risk materialised into a financial disaster when, one after another, major world banks (including some of the most powerful and influential TNCs) experienced near-total economic collapse in 2008. The cascading effects of the global credit crunch highlighted all too well how challenges run directly alongside the considerable opportunities that interdependency brings (see Chapter 9).

Activity 3

Overseas investment by TNCs is a key economic factor causing globalisation. Foreign intervention by the US government is a key political factor. Write an account discussing how the political power of the USA and the economic might of its TNCs have worked together to make this nation the leading global superpower.

Financial flows and the development process

Richer nations can assist in the economic and social development of the world's poorest nations in several ways. Four significant exchange mechanisms for directly transferring wealth from rich countries to poverty-stricken places are: (1) FDI by TNCs, (2) loan packages and (3) international aid and (4) remittances (Figure 3.5).

Figure 3.5 Financial flows and the development process

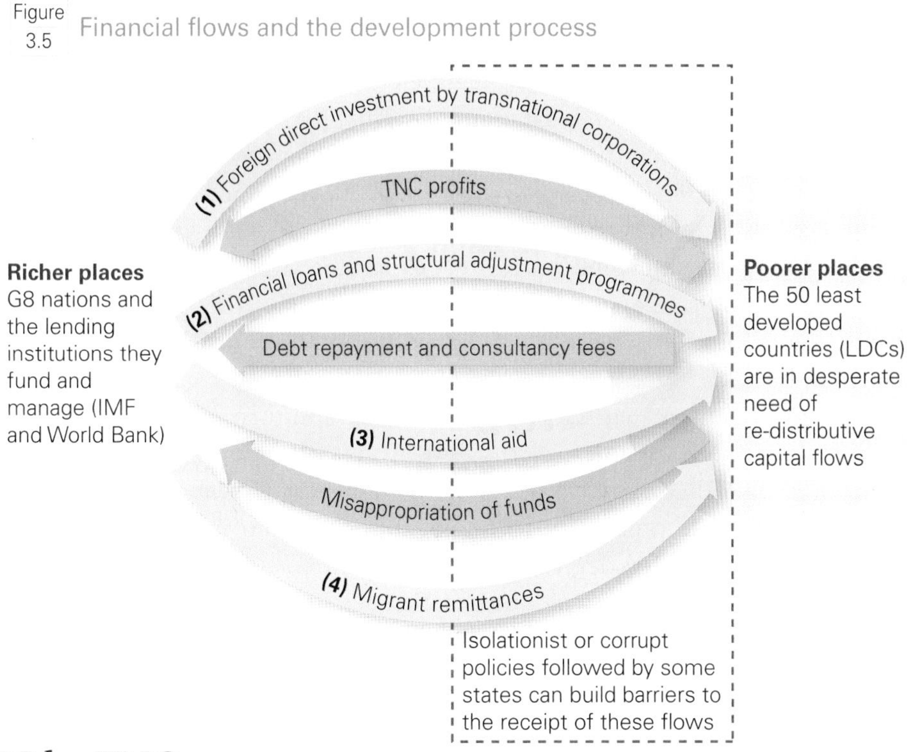

FDI by TNCs

Given that the value of foreign investment by TNCs eclipses annual figures for international aid, FDI is regarded in some quarters as the most powerful developmental tool available. The 'economic miracle' of southeast Asia's 'Tiger' economies (Hong Kong, Singapore, Taiwan and South Korea) is often held up as a shining example of the sheer transformative potential of FDI. In recent decades, these territories have enjoyed runaway industrial growth, while average wages and purchasing power are now impressive by Western standards. There can be little argument that the economic take-off of this region owes a great deal to foreign investment by Japanese and American TNCs.

However, FDI is far from being a panacea for all global poverty. Investment is disproportionately attracted to nations where human resources and site factors offer investors a good return on their money. The Tigers, advantageously located around the Pacific Rim, have these qualities in abundance — hence their recent emergence as new global hubs (Figure 3.6). Other places — notably the war-torn or landlocked sub-Saharan African countries — fare less well in soliciting attention from TNCs (Table 3.3). Finally, it must be noted that doubts linger about the real effectiveness of FDI as a developmental stimulus to recipient countries, if most of the profits are repatriated.

Figure 3.6 How natural resources and human resources help global hubs develop

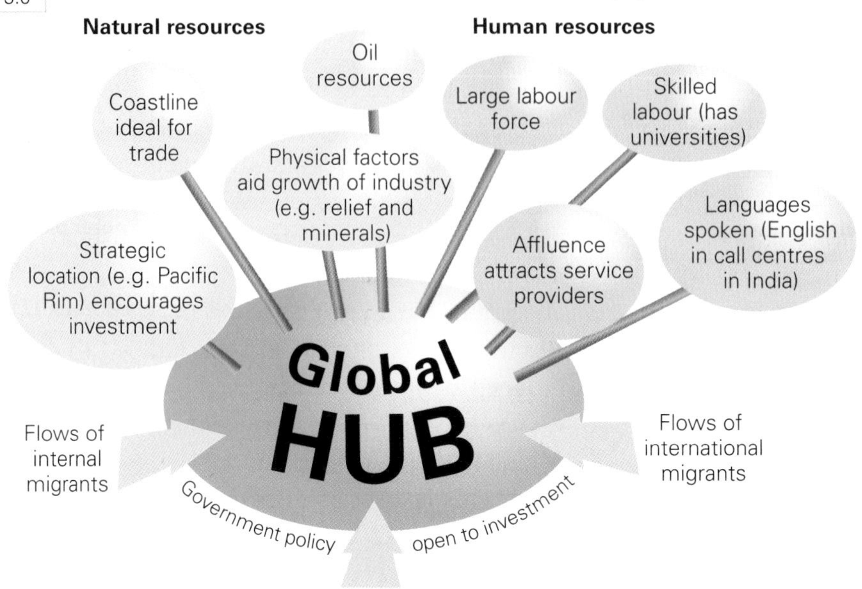

Table 3.3 Poorly connected nations: political choice or physical constraint?

North Korea	This communist dictatorship has antagonised the USA for decades. International trade is severely restricted, but since the 1990s the country has relied heavily on international aid to feed its population.	**Politically isolated**
Myanmar (Burma)	Western travel firms have ceased operations in Myanmar and the UK government advises firms not to invest there due to human rights abuses by the military junta which seized power in 1962. It is one of the most repressed, reclusive nations on earth.	**Politically repressive**
Zimbabwe	Zimbabwe's economy has collapsed under the leadership of Robert Mugabe. GDP per capita has halved since 1970 as Zimbabwe is shunned by the outside world. Investors are deterred by land seizures and election-rigging.	**Politically unstable**
Democratic Republic of Congo (DRC)	Conscription of child soldiers has interrupted the education of an entire generation; perceived as lacking in human resources and highly corrupt, the war-torn DRC will struggle to attract TNCs.	**War-torn people**
Burkina Faso and Mali	Simple lack of coastline makes these African countries unlikely candidates for outsourcing — although if education were better and the internet in place, might alternative work be found?	**Physically isolated**

Loans

Loans are one alternative means of channelling financial resources into poorer locations, usually in conjunction with FDI. Sums borrowed from the IMF and World Bank run into billions of dollars. In theory, borrowing brings prosperity to developing economies if the money is wisely invested in ways that earn an income for both the borrower and the lender (who charges interest). Indebtedness is not in itself a bad thing: rich countries routinely borrow money, while every household mortgage is a debt that usually, over time, benefits both the homeowner and the lending bank.

However, there are risks here too. For example, when interest repayment rates soared in the late 1970s and early 1980s, Mexico threatened to default on US$80 billion of accumulated debt; only emergency intervention from the IMF stabilised the situation. Since then, stringent conditions have been applied and all large global loans are accompanied by a **structural adjustment programme** (SAP). This is a set of rules designed to help avoid financial mismanagement by encouraging fiscal prudence.

Critics of SAPs point to what happened in Tanzania, where poor city-dwellers were left without free water supplies after the government was forced to privatise water supplies as a condition of its new US$143 million loan in 2003. SAPs additionally generate large return money flows into the pockets of European and American financial consultancy firms. Firms such as Arthur Andersen and KPMG typically give costly advice to poor countries on any planned infrastructure improvements. Finally, in the event of corrupt politicians embezzling money, a poor country's population will still be required to somehow repay the loan and any accrued interest.

International aid

International aid is the third major financial flow. It is a movement of money and resources made possible essentially by altruism — especially when channelled through NGOs such as Oxfam and major fund-raising events like Live Aid. The gifting of vital infrastructure such as hand-pumps or lined wells can bring real improvement in quality of life for local communities. On a larger scale, funds can be channelled into key infrastructure improvements — for instance, German finance has aided construction of Malaysia's Bakun dam. However, international aid flows have never yet been mobilised on the scale needed to transform the long-term destiny of an entire nation.

Supporters of globalisation therefore believe that TNCs remain the best hope for the future development of poverty-stricken states. Critics

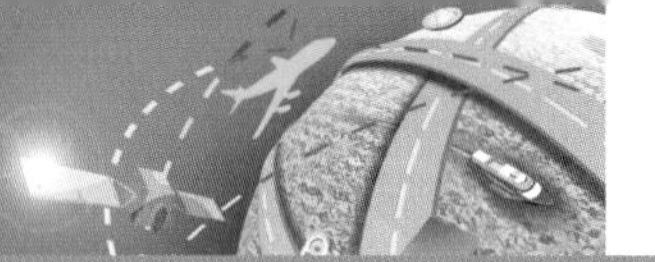

of globalisation, on the other hand, for reasons that are explained in Chapter 4, view FDI as at best a mixed blessing.

Activity 4

Identify financial flows that link you personally with the development process for poorer countries. You might consider the money paid for the clothes you wear (where were they made?), donations you may have made to charities or money spent abroad while on holiday.

4 Impacts and ethics of global interactions

Two-speed development

Moral and ethical questions are frequently raised in studies of global interaction and change. The repeated criticism is that the rich get richer while the poor get poorer as a result of globalisation. But what does 'poorer' mean in this context? For instance, the number of UK citizens living in what is termed **relative poverty** rose steeply from 13% to 22% between 1979 and 2009. Yet this figure describes the proportion of people living in households where income is less than 60% of the national average — a figure that leapt upwards in the 1980s and 1990s partly on account of tax breaks and ballooning boardroom pay and bonuses giving an enormous boost to high earners. The result was increasing inequality: a growing proportion of 'poorer' families — many of whose incomes were also improving, but less quickly — failing to keep up with a moving poverty line tied to a more rapidly increasing average wage.

Or does 'the poor get poorer' suggest a less subtle and more worrying state of **underdevelopment** — indicating instead that global interactions depress the quality of life for many people around the world to *below* what it used to be in real terms? Examples of such pauperisation can certainly be found. Most commentators agree that the **absolute poverty** rate, measured by the World Bank as people living on US$1.25 a day income or less, has increased for parts of Africa due in part to crashing primary commodity prices (coffee recently reached a 30-year low, while banana prices halved between 2007 and 2009). High rates of population growth and economic instability on account of prolonged conflict have also played a role in depressing incomes, for example in the DRC and Uganda.

In contrast, however, numbers living on extremely low incomes decreased throughout Asia over roughly the same time period (Figure 4.1). Improvements in life expectancy accompanied economic progress across the region (Table 4.1). Most significantly, 400 million Chinese were lifted out of absolute poverty between 1979 and 2004. This represents the greatest single leap forward in poverty reduction the world has ever seen and corresponds

Figure 4.1 Changes in the world distribution of annual income (WDI)

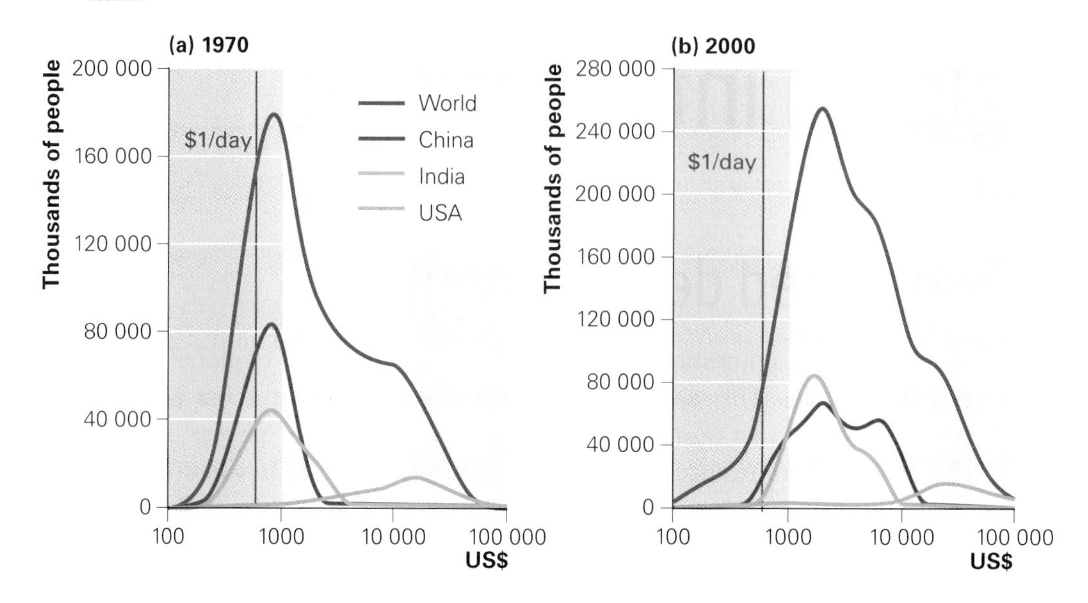

neatly with China's period of economic reform and increased global interaction. Of course, many Chinese continue to exist on meagre wages by Western standards and a rural rump of just under 100 million people still earn less than 637 yuan (£43) a year. It is the growing development gap between their own rich and poor that now concerns China's rulers most: the divide has never been wider than today, thanks to runaway success for the country's new economic elite. China is a 'two-speed world' in its own right, while simultaneously being part of a larger, constantly changing picture of global inequality (Figure 4.2).

Thus, while the image of a two-speed world roughly grouped into bundles of rich countries and poor countries may still be loosely accurate, it is now also the case that global interactions are responsible for a growing wealth divide *within* nations (Table 4.2). In India and Indonesia, the majority of people are economically worse off than ever before in relation to the richest members of society — even if that same majority are generally better off than previous generations when their income is measured in real terms.

Table 4.1 Changing life expectancy in parts of Asia

Country	Life expectancy at birth (years)	
	1950	2000
Bangladesh	36.6	58.1
Bhutan	35.2	60.7
North Korea	49.0	63.1
India	38.7	62.3
Indonesia	37.5	65.1
World	**46.5**	**65.0**

Figure 4.2 A two-speed world: the planet today in numbers

Poverty in sub-Saharan Africa remains at **50%** — no lower than in 1981; and population growth has meant the number of poor living here has doubled, from 200 million to about 380 million	Poverty in east Asia fell from 80% of the population living below US$1.25 a day in 1981 to 18% in 2005; much of the progress was in China, where **0.4 billion** have been lifted out of absolute poverty	Of the half a billion people worldwide who escaped **US$1.25**-a-day absolute poverty during 1981–2005, most would still be deemed very poor by European and North American standards
The richest **seven** people alive today hold wealth equivalent to the GDP (gross domestic product) of the 41 highly indebted poor countries (HIPC) — home to 567 million people	Division of the world's **US$50 trillion** wealth between haves and have-nots creates **groups** of rich and poor countries; it also **cuts across** all countries where a rich elite co-exists with a poverty-ridden underclass	**1.4 billion** people live on less than US$1.25 a day, the World Bank poverty measure; and at least 80% of humanity lives on less than US$10 a day (purchasing power equivalent)
Half a billion Indians lived in US$1.25 poverty in 2009; yet 24 of the 800 richest people on the planet were also Indian citizens — including **two** of the **top-ten** earners	The world's richest nations are also home to 100 million people who live below these places' official poverty line; this figure includes **37 million** American citizens	Global inequality is worsening: in 1960, the 20% of the world's people who live in the richest countries had 30 times the income of the poorest 20% — in 1997, **74 times** as much

Table 4.2 Income inequalities between and within selected nations, 2005

HDI Rank		GDP (current US$ billions)	GDP per capita (US$ thousands)	Share of national income going to poorest fifth of population (%)	Share of national income going to richest tenth of population (%)
High human development					
12	USA	12416	42	5	30
16	UK	2198	37	6	29
Medium human development					
107	Indonesia	287	1.3	8	29
128	India	805	0.7	8	31
Low human development					
159	Tanzania	12	0.3	7	27
165	Zambia	7	0.6	4	39

In summary, during the era of modern globalisation:

- Worldwide absolute poverty has fallen — but this bland statement hides the extremely uneven performance of different world regions.
- Many countries have, as a whole, advanced from low-income to middle-income status since the 1970s — although the development gap between the very richest and poorest groups of nations is enormous and widening.
- Within-country inequality has increased across the board.

Activity 1

Analyse Figure 4.1, taking careful note of the logarithmic scale. In your answer, describe (a) changes shown over time in the proportion of people with extremely low incomes and (b) the changing ranges of incomes. Provide a summary statement (what do the data tell us about the changing nature of global income inequality?)

Core–periphery structures

The core–periphery models of Myrdal, Hirschman and Friedmann can be taken as a starting point for the investigation of 'two-speed' geography. Core regions — on a global or regional scale — are places that enjoy cumulative growth processes fuelled by flows of raw materials, migrants and entrepreneurial talent from surrounding peripheral areas (Figure 1.8, page 15). This process of uneven development — the spatial movement and concentration of physical and human resources into a core — is called **backwash**. In time, growth is predicted to spread into the periphery as a result of market recompense for raw materials, the diffusion of innovations from core regions, government intervention and other beneficial interactions (Figure 3.5, page 43). In Friedmann's account, a more complex core–periphery system develops over time as a result of spreading wealth. At first, secondary cores materialise in the periphery; later, a functionally interdependent system of linked core regions evolves (Figure 4.3).

The same optimistic trickle-down view of spatial economics has been a hallmark of Washington–IMF monetary policy-making since the 1980s. And, at first glance, recent global economic changes might be described and explained in similar terms. Thus, the emerging economies of Asia, their growth stimulated in part by FDI trickle-down, are cast as new secondary cores, now rising in power and influence. Yet even with this refinement, core–periphery model-making gives an incomplete picture of global wealth distribution. This is because money is now being channelled into the pockets of a global elite who are no longer spatially contained within the borders of

Figure 4.3 Friedmann's core–periphery model showing the growth of functional interdependency over time

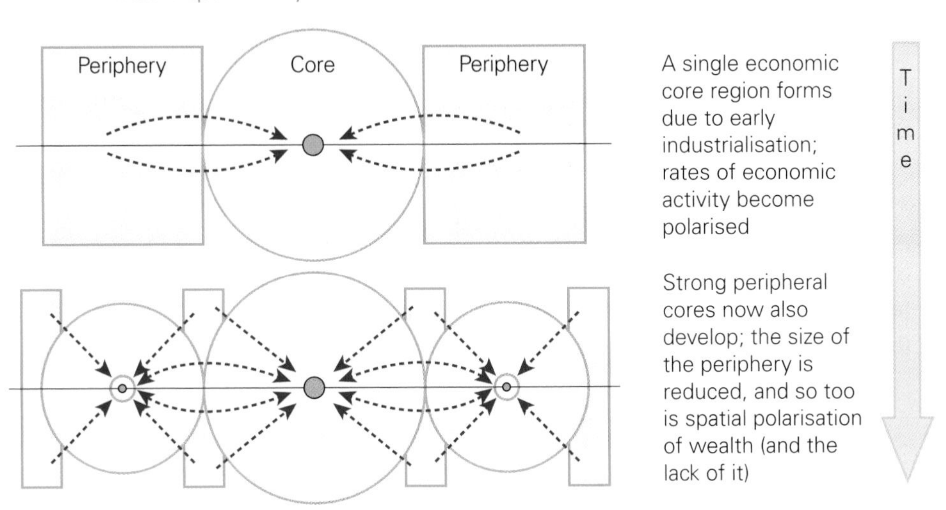

a mere handful of countries. The world billionaire map provides far better insight into the increasingly diffuse spread of extremely rich people living in today's two-speed world (Figure 4.4).

Figure 4.4 The global spread of the world's economic core group: the number and distribution of billionaires in 2009

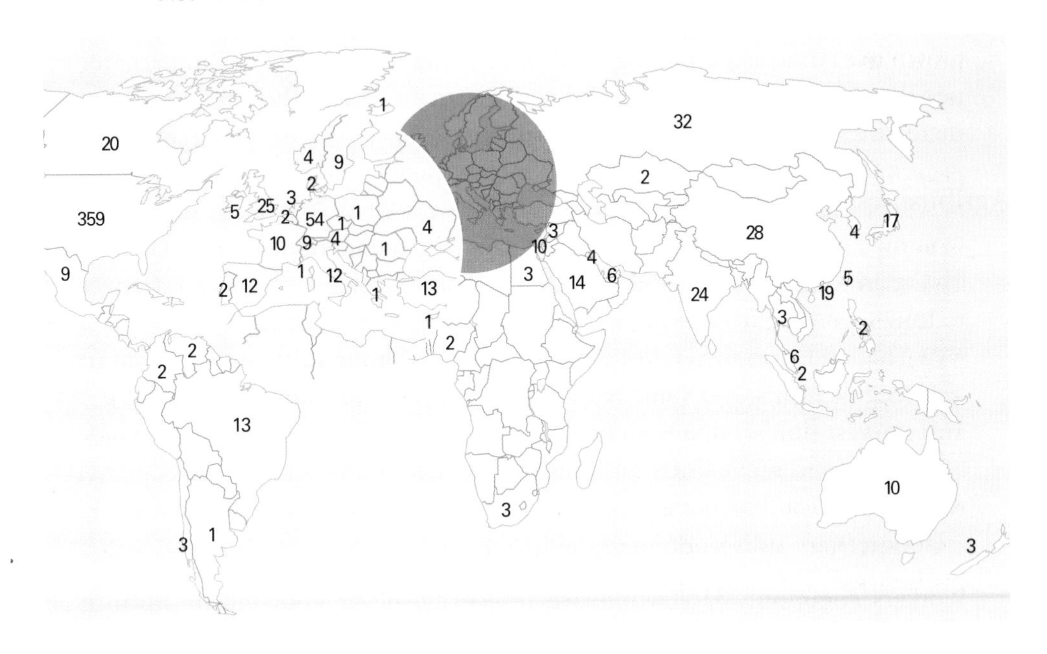

Activity 2

Visit **http://hdrstats.undp.org/buildtables/**. Conducting research using this important UN website will help you develop independent study skills. Using the UN statistics, you can design tabulated datasets to display income gaps and other developmental indicators for carefully selected nations. (Figure 4.4 was produced using this resource. Start by attempting to replicate it.)

Activity 3

Study Figure 4.4.

(a) Describe the global distribution pattern of billionaires.

(b) Consider to what extent this pattern reflects national differences in levels of average wealth.

Global shift and the changing nature of work

According to geographer Peter Dicken, the global pattern of work has evolved well beyond the traditional division of labour where farmers and miners in peripheral developing countries supplied raw materials for the factory workers of global core regions. More complex networks of economic activity have grown over time as part of a **global shift**, bringing wide-ranging impacts for people in low-income and middle-income nations — whose employment increasingly straddles all sectors of industry.

Agribusiness workers

Commercial agriculture is a major global employment provider dominated by TNC giants such as Del Monte and Cargill. This industrial sector is no stranger to tough working conditions, as the following examples show:

- In Thailand, commercial rice farmers work 12-hour shifts bent double in 38-degree heat — yet some receive wages of just US$2 a day (Figure 4.5).
- At harvest time, Indonesia's prawn aquaculture labourers sometimes work both day and night shifts on inhospitable tidal mud flats.
- In Costa Rica, banana crops are sprayed with hazardous pesticides while labourers are still working in the fields.

Workers commonly face workplace insecurity, never knowing in advance what hours they will be required for. Some large UK supermarkets send

Figure 4.5 Transplanting rice in Thailand

J Marshall — Tribaleye Images/Alamy

automated e-mails to their Kenyan vegetable suppliers requesting increased production quotas whenever store tills register especially good sales for any particular item. This 'just-in-time' ordering makes perfect financial sense for these stores; but for women with daily childcare responsibilities, continued agricultural employment becomes a near-impossibility under such irregular and unpredictable working conditions.

It is by no means the case that all globally networked agribusiness labourers are treated poorly, however. Workers who organise themselves into collectives, and partner with responsible retailers that subscribe to the ETI (ethical trading initiative) base code, find their working lives greatly improved. South African wine producers supplying UK supermarket Waitrose provide one such example.

The manufacturing sector

In Chapter 3 we explored reasons why manufacturing codes of conduct (health and safety) are often poorly enforced. As a result of such weak regulation, 2500 metal workers in the Chinese city of Yongkang lose a limb or finger each year while assembling goods under contract for firms which are household names like Black & Decker. In Dongguan, workplace insecurity haunts workers with families to feed. Temporary contracts are the norm in this Chinese city, dubbed 'Santa's workshop', whose economy is driven by intensive production spurts in advance of Christmas orders for Western stores.

That workers continue to tolerate such insecure conditions is symptomatic of a fundamental lack of bargaining power with employers. This is due to factors such as:

- unemployed labour waiting in the wings to replace disobedient employees
- low tolerance of trade unions by governments fearing 'capital flight' by TNCs

- the common practice of withholding wages owed in arrears in event of an employee quitting

In one extreme case, female textile workers in Nicaragua testified that they were sacked for joining unions; some claimed promotion could be gained from management only in exchange for sexual favours.

Of course, not all new work environments fostered by the global shift of manufacturing are so unappealing. One young Chinese worker became an overnight global celebrity in 2008 when she inadvertently left a photograph of herself stored on a brand new iPhone following factory testing. The image shows a cheerful smiling girl, who probably makes around US$100 a month, working in a clean, hi-tech environment for the Taiwanese company FoxConn (which manufactures Apple iPhones in Shenzhen). The picture was duly discovered and posted online by the purchaser (see website: http://forums.macrumors.com/showthread.php?p=6076473/).

Call centres and tertiary work

Chapter 2 explored how broadband internet has allowed the offshoring of white-collar work in addition to blue-collar factory employment. New opportunity has transformed lives in Bangalore, India's fastest growing city with a global reputation for call-centre telephone inquiry work. Broadband capacity is unusually high in this long-established technology hub, thanks to investment in the 1980s by domestic media companies working alongside

Figure 4.6 Call-centre workers in Egypt

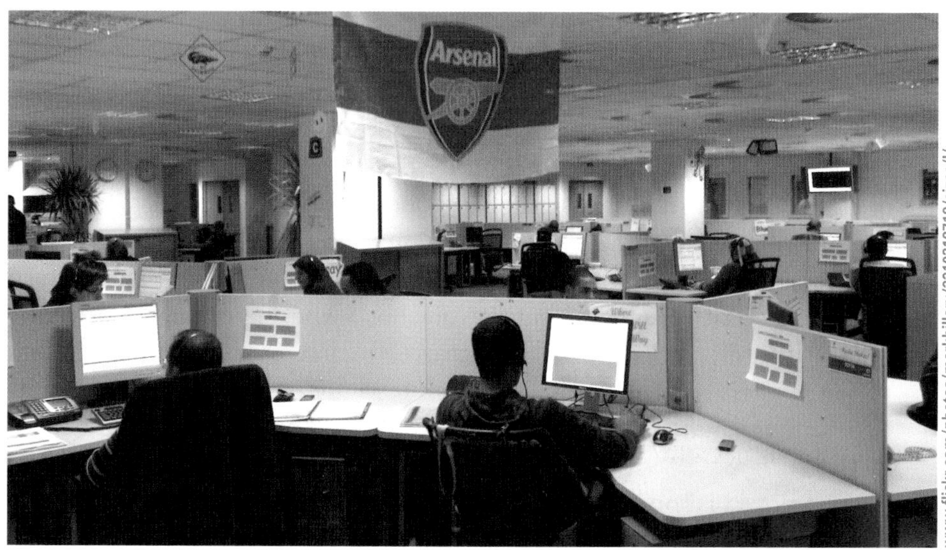

www.flickr.com/photos/markhillary/3310318718/sizes/l/

foreign firms like Texas Instruments. Dell, Intel and Yahoo have since established their own call centres here, while large independent operators now conduct contract work for all kinds of firms, from travel companies to credit card providers.

Nightclubs and 24-hour shopping malls in Bangalore testify to the relatively high purchasing power of a new Indian 'techno-elite' typically earning 3000 rupees (£38) a week. The work can be demanding, however:

- Many of the calls answered are low-level transactional enquiries, rendering employment repetitive and unenjoyable.
- Business is often conducted at night, due to time zone differences between India and customer locations in the USA or UK, and sometimes in 10-hour shifts, 6 days a week.
- Some employers demand that workers adopt an artificial Westernised identity (such as an anglicised name), minimise local features of their speech and conceal the call-centre location from customers — all of which may cause psychological tension for staff.

Yet compared with the lives of hundreds of millions of rural Indian labourers (working equally long hours but for as little as 250 rupees a week), call-centre workers enjoy superior employment, both in Asia and other continents, including Africa (Figure 4.6).

Quaternary research work

Research and development in information technology, biotechnology and medical science comes under the umbrella of quaternary-sector work. The high skill level demanded by this fourth type of employment has not prevented its global shift, and in rising numbers too. Across China, India and South Korea, 2500 universities produce millions of new graduates each year. China alone awarded 25 000 PhDs in 2007, three-quarters of them in science and engineering. South American and South African universities boast impressive research clusters, stimulating home-grown research while also conducting outsourced investigative work for foreign industries.

- In China, US firm Intel employs 1000 research staff on the outskirts of Shanghai. ABB, the Swedish–Swiss engineering group, has a research wing in Beijing, as do Nokia and Vodafone.
- Southeast Asia is a global hub for stem-cell research. South Korea's government allows work to be freely conducted using human embryos, which has been limited by legislation in the USA.

Medical research is increasingly conducted overseas by Western firms such as Pfizer and GlaxoSmithKline. Some 700 clinical trials were conducted in India, China and Russia during 2008, up from fewer than 50 in 2003. International welfare organisations are concerned that poorly paid volunteers for drug trials in these countries are being exploited, serving as guinea pigs to test medicines they may never be able to afford. Here too in the quaternary sector, global shift can seem ethically ambiguous and morally questionable.

Evaluating and managing the social impacts

The UN Millennium Development Goals set global poverty-reduction targets for 2015 (Figure 4.7). Global interaction both helps and hinders this endeavour — depending on the nature of the work involved and levels of remuneration and regulation. Although the shift to salaried employment and urban living has lifted many millions out of rural poverty, 80% of the world's population still lives on less than US$10 a day. Meanwhile, the greatest beneficiaries of globalisation have acquired mind-blowing levels of personal wealth. The global financial system produces and reproduces unevenness on an epic scale.

Opportunities for ethical intervention do exist, though, and at varying scales of action (Figure 4.8). Power to effect real change is distributed throughout global networks — from producers (who can seek new ways of working with TNCs) to consumers (without whose goodwill all businesses must fail). Sometimes the terms of globally shifted trade and employment have been re-negotiated to deliver a more equitable outcome for people at the bottom of the supply chain — as, for instance, whenever a coordinated effort of producers, suppliers and consumers builds a new Fair Trade franchise.

Figure 4.7 The Millennium Development Goals

The Millennium Development Goals (MDGs) are eight specific goals to be met by 2015 that aim to combat extreme poverty across the world:

- Eradicate extreme poverty and hunger.
- Achieve universal primary education.
- Promote gender equality and empower women.
- Reduce child mortality.
- Improve maternal health.
- Combat HIV and AIDS, malaria and other diseases.
- Ensure environmental sustainability.
- Develop a global partnership for development.

These goals were agreed upon at the UN Millennium Summit in New York in 2000. The declaration was adopted by 189 nations and signed by 147 heads of state.

Figure 4.8 Trade networks and power

Factory labourers can seek solidarity and may fight for their political right to create trade unions or negotiate a minimum wage and other benefits

TNCs can buy from workers' cooperatives or source goods ethically — they may do more to enforce codes of conduct on their own networks of **suppliers**

Farm workers can organise themselves into collectives and may attempt to re-negotiate terms of trade with suppliers, especially Fair Trade organisations

Producers and consumers are linked with other actors in other **places** and at different **scales**. The **power** to act — and to effect **change** — is embedded in many different locations within the network; the most effective changes are often brought by different actors or places working together in **partnership**

National governments could do more to regulate the TNCs domiciled in their countries. **Supranational organisations** such as the EU or the WTO might reform rules regulating global trade

NGOs and **charities** can lobby, raise public awareness and fund projects. **Educational courses and materials** — including this book — could have a role

Consumers are moral beings who may ask questions about the other humans they are linked with in supply chains; they can knowingly reject exploitative goods

Activity 4

Find your own 'location' in Figure 4.8. Identify other players that you can become connected with through your own personal actions — such as purchasing goods or supporting a charity. Discuss whether you would consider yourself to be one of the more powerful or less powerful players shown in Figure 4.8.

Action has also been required to assist less well-off people in richer countries who have lost jobs as a result of global shift. Parallel to the growth in ship-building and steel-making employment in nations like South Korea, British cities such as Glasgow and Sheffield saw their local economies decimated when the first waves of redundancies in those industrial sectors hit home during the 1960s and 1970s. A **negative multiplier effect**, or vicious circle of decline, ripped though city services and retailing; further symptoms included industrial militancy, civil unrest and the onset of a vicious cycle of poverty (Figure 4.9). Similarly in the USA, inner-city communities in Baltimore and Detroit found themselves suddenly 'switched off' from the wider world economy (though not always from the informal global interactions that comprise drug-trafficking). Many European and American manual workers, almost overnight, became irrelevant to the world economy, abruptly ceasing to be significant producers or consumers of wealth.

Figure 4.9 A deindustrialisation timeline for the UK

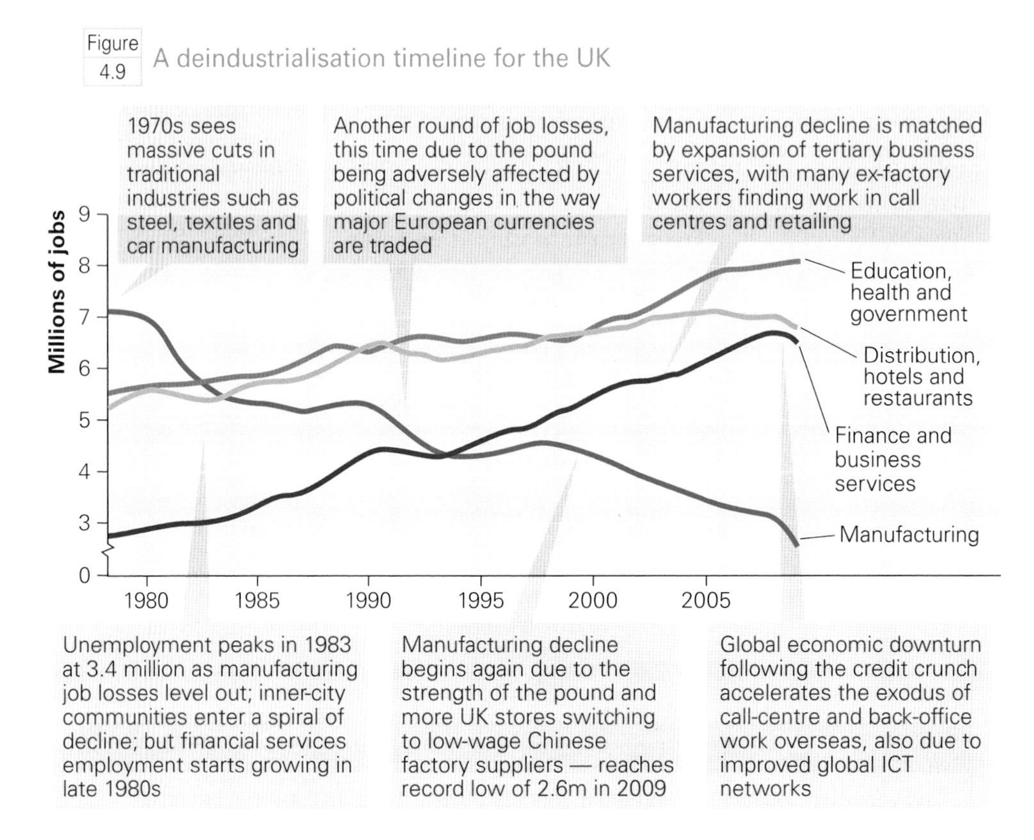

Today's TNC management is more mindful of what is now termed 'corporate social responsibility'; senior staff may increasingly be observed making some effort to navigate the world with a moral compass. For instance, design entrepreneur James Dyson, famous for his bagless vacuum cleaners, resisted the relocation of his own firm's manufacturing wing to Asia for as long as he reasonably could, in order to safeguard employment of his English workforce. But in 2003, profitability concerns finally saw the Dyson Corporation shop floor moved to Malaysia. Even today, no firm is likely to occupy an unprofitable production site for long when shareholder interest is at risk, despite the unfortunate social effects of factory closure. So global shift remains an ongoing affair, now supplemented by an ever-increasing volume of office work, enabled by broadband — and driven by the never-ending need to trim operating costs in an increasingly competitive economic environment.

5 Cultural globalisation

Cultural interactions and impacts

Definitions of culture typically refer to the everyday or common values and traditions of a group of people. Yet such things do not remain fixed: they are repeatedly renewed, recreated and modified over time. Thus, any discussion of the global spread of an 'English culture' is complicated by the fact that the **cultural traits** (views, tastes, fashions, beliefs and speech patterns) of ordinary people in England are already very different from 100 years ago, in part due to global influences (Table 5.1). This fluid and constantly shifting character makes culture a challenging topic for geographical analysis.

Many important writers have analysed how culture can develop and change over time, including Raymond Williams and Stuart Hall. Central to

Table 5.1 The English people's changing cultural traits

	Early twentieth century	Twenty-first century
Religious beliefs	Generally widespread, with high levels of church attendance	Largely secular and non-religious although some minority faiths and beliefs are prospering
Food	Locally sourced seasonal food, rarely using strong foreign spices, preferring native herbs	Global, varied tastes in food. Strong spices are widely used in cooking
Identity	People have a strong sense of local belonging (either to a town or county). Regional dialects are stronger than today; most are also extremely patriotic and would fight for their country	Many would be less willing to fight for their country, although they are often strong supporters of national football teams. Younger people may see themselves as 'global citizens'
Roots of vocabulary	Anglo-Saxon, Norse, French, Latin, Greek and others	Additional influences, e.g. Indian, Jamaican, American English
Entertainment	Oral story-telling (low literacy), religious hymns, folk music and local music halls	Modern popular music performed by global acts. Huge importance of soaps and reality television

their work are two key concepts: ideology and power. An **ideology** is a set of beliefs belonging to a particular group of people. It may begin as a minority viewpoint; most of the world's major religious and political manifestos began with just a handful of followers. Over time, these beliefs came to be shared by far greater numbers. The rise of fascism in Germany during the 1930s is perhaps the most well-known example of a fast-moving and mobile new ideology quickly 'converting' ordinary people to an unconscionable cause — transforming local cultures, not just in Germany itself but around the world. Nazism was a very powerful ideology.

Analysis of cultural globalisation begins by asking: why are some cultures and ideologies especially effective at building global influence? Successful cultural traits and beliefs travel in varied ways:

- International migration brings direct mixing of cultures.
- Tourism brings direct contact between people of different cultures.
- Hollywood films have been viewed by large overseas audiences since the early 1900s.
- Internet chatrooms and virtual communities foster personalised contact between people living in different countries.
- TNC stores, especially fast-food and clothing chains, influence tastes in emerging world markets through the introduction of new products.
- TNC factories and offices can bring behavioural and linguistic change to local people working there.

The global diffusion of 'Western' cultural influences — notably American and European food, fashion and a consumerist lifestyle — is achieved through use of all these different channels. The USA exerts a strong cultural influence over other nations that goes hand in glove with its economic power. Enormous American TNCs such as Disney and Coca-Cola have the financial clout to devise seductive advertising for different regional markets, 'enrolling' local people into their global networks of consumption.

Is this a good thing? Critics argue that the 'Westernisation' of global culture is an echo of nineteenth-century colonialism. Yet supporters of globalisation (sometimes called the 'hyperglobalisers') see such comparisons as unduly negative. Western music and Hollywood films, they argue, can also serve as a 'carrier signal' for important progressive cultural messages. An empowered American woman such as Madonna may serve as a role model for young women living in societies where men still treat them as second-class citizens. People living under dictatorships learn about democracy through exposure to films and websites produced in cultures where universal suffrage is a fundamental human right (Table 5.2).

Table 5.2 Different views on cultural globalisation

Hyperglobalisers' view	Transformationalists' view	Sceptics' view
Globalisation will reduce the relevance and power of countries and will result in positive cultural exchanges. This will, over time, reduce group attachments to ethnic and religious identity. It will therefore increase global and local prospects for peace — by reducing possible opportunities for prejudice to arise in a global village.	Globalisation brings a threat of 'other' influences to places, often prompting fascistic reactions and the reassertion of national identity by a right-wing 'core' group who feel threatened. But transformationalists believe this new challenge is balanced out by the fresh opportunities that globalisation brings to different cultures.	Globalisation is a process that allows the continued dominance of powerful Western regions. 'Cultural imperialism' is part of this project: an exporting of Western ideas helping to change the values of people around the world — also making them believe that free-market economics and rampant consumerism is the only way forward.

Positive → **Negative**

However, globalising ideologies are sometimes contested and resisted rather than being accepted and absorbed. Cultural globalisation is not a simple process, and a spectrum of local responses can be observed. In 2002, militant Islamist group Jemaah Islamiyah bombed Bali tourist resorts popular with holidaymakers from North America, Europe and Australia. In part, the intention was to send a message deterring Western tourists from travelling to Indonesia.

Activity 1

(a) For the country where you live, write down a list of ten things that you feel are strongly representative of the shared national culture. The list could include a mixture of famous landmarks, songs, important traditions, food items or other important 'signifiers' of culture. For instance, a list compiled for England might be quite likely to include tea-drinking — or perhaps Stonehenge? Compare your response with those of other students: how easy is it to define what an English, Welsh, French etc. 'culture' is?

(b) If you intend to research this topic further, key enquiry questions are: why do some people living in certain places resist foreign cultural influences, whereas others accept them? What kinds of power relations are at play?

Globalisation and local culture

The guiding principle of ambitious firms is to keep building market share, preferably on a global scale. The continued success of major Western music and media providers, or fast-food franchises like Pizza Hut, depends on cultural acceptance and approval of their products. Logically, such firms might be seen to have a vested interest in using advertising to persuade more people in Asia, Africa and Latin America to abandon their own traditions and instead embrace European and North American tastes. In theory, capitalism has much to gain from fostering such **cultural homogenisation** on a worldwide scale — in order to maximise profits more readily. Following this logic, a world where everyone enjoys beefburgers would surely be good news for Burger King. One single global community singing the songs of Coldplay ought to delight music copyright owner EMI.

Such homogenisation does indeed sometimes occur — and for the reason described. For instance:

- Consumers using British firm Tesco's Asian supermarket stores, for example in South Korea and Thailand, now enjoy a wider range of imported dairy products than they grew up with.
- After 20 years of economic and cultural exchange with the West, some young Chinese musicians are turning their backs on traditional folk music and writing guitar-based rock songs instead.

Critics of cultural globalisation describe such transformations as the Americanisation or McDonaldisation of our world. The largest Anglo-American TNCs are indeed experts in designing and advertising aspirational products (Figure 8.2, page 100). The most successful appear able to enrol as new consumers people belonging to a whole range of different local cultures. Notable global 'success' stories include the following:

- **Walt Disney Company** This is the world's largest entertainment TNC with annual earnings of US$38 billion. Across Asia, young children encounter Mickey Mouse and similar brands on Disney Channel Asia (Figure 5.1).
- **Coca-Cola** The sugary drink manufacturer also has a global presence (diffusion of Coca-Cola into new markets during the Second World War and later in the 1950s correlates with patterns of US military activity at that time).

The powerful cultural influence exerted by economic agents of the world's richest nations is exhibited in other ways too. The continuing loss of indigenous languages is one particular cause for concern. The English tongue has supplanted native languages in North America, while Spanish and Portuguese have had similar effects in Latin America. Under direct colonial rule this was often to be expected; in today's postcolonial era, powerful global media providers, the internet and other factors continue to exert an indirect influence on language that threatens further loss of local cultural diversity.

Figure 5.1 Disneyland, Hong Kong

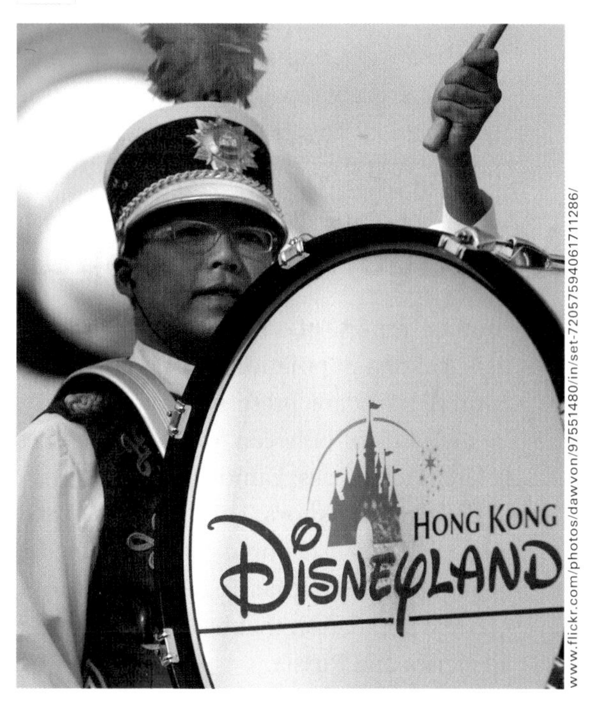

However, reduced cultural diversity is not an inevitable result of the global actions of TNCs and other powerful players. Actual outcomes — as experienced at the local level — may be far more complex than the Americanisation hypothesis suggests. In truth, Anglo-American TNCs probably do not have a hidden agenda of cultural imperialism and have not purposely set out to destroy local cultures. The opposite may even be the case. New ideas taken from the world's myriad local cultures drive innovation in creative industries. Film, music and food industries have all thrived by mixing together Asian, South American and African influences with European and American ideas. The evidence also suggests that globalisation creates exciting new mixed, hybrid cultures. We will learn more about this in Chapter 8.

Activity 2

Conduct online research for your own example of a TNC (other than Disney, Coca-Cola or MTV) that you associate strongly with global 'Americanisation'. (Marvel Entertainment and Burger King are two possible suggestions.)

(a) Try to identify how many countries your TNC sells its products to.

(b) Look for important data such as annual profits or workforce statistics.

(c) Is there any evidence of your TNC adapting its products for the varying tastes of different markets?

Activity 3

Conduct a 'language audit' for your class, or your entire year group and teachers. How many languages are spoken? Which are the most common? Are there any unusual languages spoken by just one or two people? How many people speak only one language? What is the highest number of languages spoken by a single person?

Migration and diaspora

Another aspect of cultural globalisation is the worldwide scattering or dispersal of a population. This is called **diaspora**. Over time, each diaspora's cultural traits are preserved (albeit in a modified form) and connections are maintained between groups of peoples with common ancestry living in different territories. Famous examples include the following:

- **The Jewish diaspora** This first historical diaspora was accelerated after AD 70, when the Jewish population was forced to disperse following the Roman conquest of Palestine. Today, the state of Israel serves as the hub for a global Jewish population that has notable concentrations in western Europe, North America and Russia.
- **The 'black Atlantic' diaspora** This has been described by writer Paul Gilroy as a 'transnational culture' built on the movements of people of African descent to the Caribbean, the Americas and Europe. A shared, spatially dislocated history of slavery originally helped shape this group's identity. Today, international connectivity is maintained through tourism and cultural exchanges across the Atlantic, exemplified by an international black music scene that has given the world jazz, Jimi Hendrix, reggae and hip-hop.
- **The new Polish diaspora** Some 1–3 million young people left Poland after it joined the European Union in 2004. These recent émigrés join a larger and more well-established mass of around 18 million people of Polish descent living abroad, some of whom are the descendants of Second World War exiles.
- **The Chinese diaspora** Indonesia, Thailand, Malaysia and other southeast Asian countries, along with far-flung places such as the UK and France, have significant Chinese populations (Figure 5.2). In many world cities, clearly delimited 'Chinatown' districts exist. A thousand years of seafaring trade gives this diaspora a long history, and Chinese is the world's most spoken language in terms of overall number of speakers. Currently, many Chinese are working for Chinese enterprises abroad, especially in Africa.
- **The 'Celtic' diaspora** The countries of the 'Celtic fringe' of the British Isles — Ireland, Scotland and Wales — have all produced significant global

diasporas despite these nations' relatively small population sizes. For instance, Ireland is home to just 4 million people, yet over 70 million individuals worldwide claim Irish ancestry. In the USA alone, 30 million people believe themselves tied to Irish bloodlines, following mass emigration from Ireland especially during the nineteenth and early twentieth centuries.

Figure 5.2 Current numbers of ethnic Chinese outside China and Taiwan

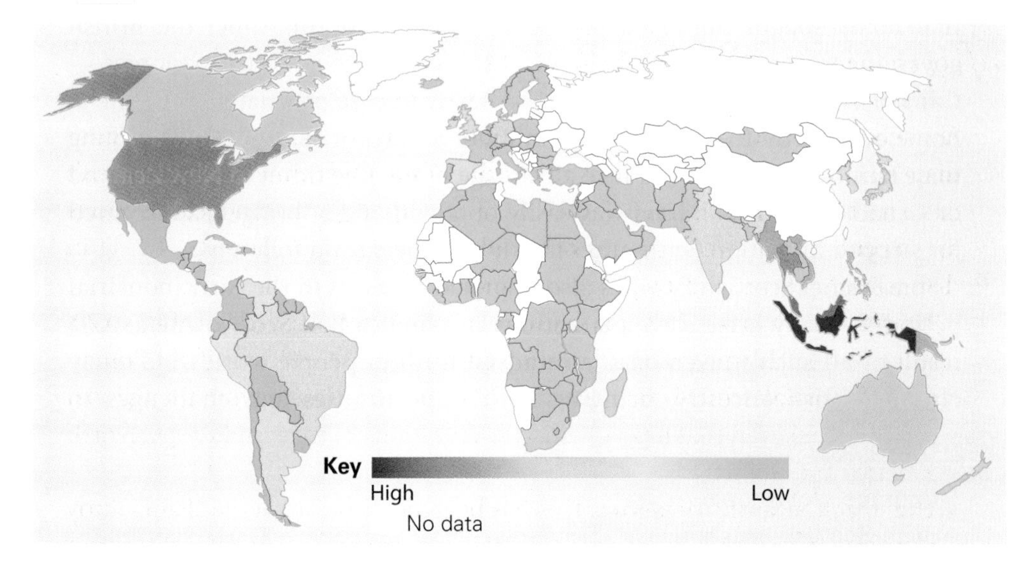

Activity 4

There are many more examples that can be researched, including French, Indian, Italian, English and Malay diasporas. An excellent example of a rapidly growing new global diaspora is the Mexican one. Using your own research, make a case study of Mexicans and people claiming Mexican descent living in the USA (and other countries such as Canada). The US census website is a rich source of information: **www.census.gov/**

Scotland: a diaspora source nation

Although its roots are old, the Scottish diaspora has thrived in the era of globalisation thanks to cheap air travel and the internet allowing a scattered population to become far better connected with one another. Worldwide dispersal of Scottish people dates back to the late eighteenth and the nineteenth centuries, when Scotland's rural poor suffered an internal and subsequently international displacement called the Highland Clearances. Prior to the Industrial Revolution, peasants were given subsistence farming ('crofting') land rights by clan chiefs. However, by the late 1700s, crofting tenancies were no longer being renewed. A shortage of land for sheep grazing pushed up agricultural rents, with the demand for wool coming from Britain's fast-growing factory towns. As a result of these changes, between 1780 and 1880 up to 150 000 Scots were evicted from their crofting homes and made landless.

The economic refugees at first relocated to marginal coastal areas of Scotland. However, the carrying capacity of this land was often so low that many sought improved prospects overseas. At this time, the British government was offering British citizens free passage to certain colonies — Canada, Australia and New Zealand — partly to ease population pressure at home and partly to shore up Britain's global territories. Typically, a young male family member would take advantage of the free ticket to New Zealand or Canada. On arrival, land was easily obtained and a 'bridgehead' secured for stepped migration, with other family members soon following.

Some richer Scottish families also migrated overseas in the early industrial period — many large slave plantations in Jamaica had Scots owners. As a result of all such movements, around 40 million people worldwide today claim Scottish ancestry, bringing both opportunities and challenges to

Table 5.3 The Scottish diaspora (numbers of people with Scottish ancestry and their location)

Britain	
Scotland	5 000 000
England and Wales	2 000 000
Rest of world	
USA	21 000 000
Australia	5 000 000
Canada	4 500 000
New Zealand	2 000 000
South Africa	1 000 000
Ireland	1 000 000
Brazil	1 000 000
Netherlands	750 000
Poland	500 000
France	500 000
Caribbean	500 000
Scandinavia	250 000

Figure 5.3 A display of 'tartanry' — is Scottish culture becoming 'fossilised'?

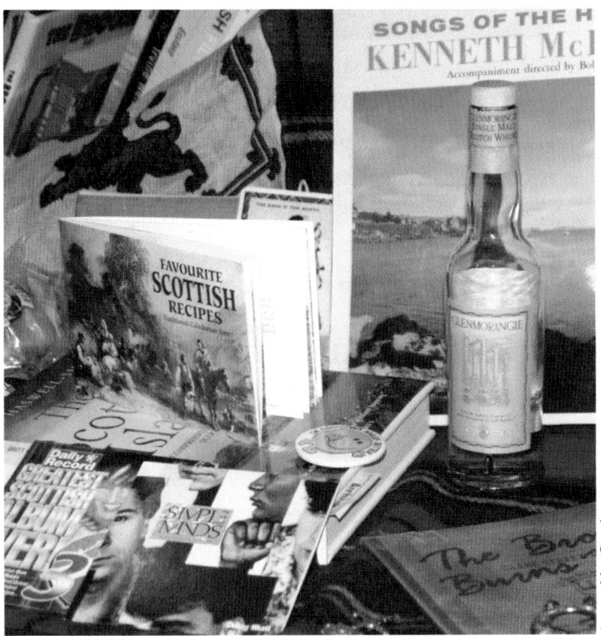

Jones-MacSwein

Scotland. Ancestry websites enable people living all over the world to trace their roots back to Scotland (Table 5.3). Many are keen to visit, making this an enormous prize for the Scottish tourism industry, which counted 2.8 million overseas visitors in 2007. GlobalScot is a website run by government-funded Scottish Enterprise that encourages members of the Scottish diaspora to economically network with one another. Diaspora communities also thrive in the virtual environments of Facebook and Twitter.

Sensing business opportunity, a large part of Scotland's tourism industry plays up to the expectations of tourists with Scottish roots who want to see 'tartan and bagpipes'. However, many young people in Scotland object to this idolisation of the past; they fear their nation risks becoming culturally fossilised (Figure 5.3).

The USA: a diaspora host nation

The Scottish diaspora is well represented in the USA, along with around 300 million other people of foreign ancestry who share the territory with just 3 million Native Americans. The original 'immigrant nation', the USA is home to people of Scottish, Irish, Italian, Greek, Jamaican, Puerto Rican, Indian, Swedish, Polish, Jewish, German, Korean, Nigerian and Chinese descent, among many others (Figure 5.4). *The Simpsons* television programme provides a contemporary comic view of US society as a home for diaspora groups, featuring characters like Scottish stereotype Groundskeeper Willie and Apu, the Indian shopkeeper.

Figure 5.4 Immigrants arriving at Ellis Island, USA, in the early twentieth century

The National Archives

The ongoing celebration of ancestral ethnic identities is a significant feature of US society. St Patrick's Day brings a major parade to cities with large Irish populations, such as Boston and New York. Many people of non-Irish origin also celebrate the day in company with their Irish-American friends: here we can observe the adoption of Irish cultural traits by wider society. Words have also become shared over time. Television shows such as *The Sopranos* (a drama about an Italian-American Mafia family) and *The Wire* (set partly in Baltimore's black ghettoes) have popularised forms of ethnic-minority speech.

There is money to be made here for entrepreneurs investing in businesses that draw on diaspora culture. Irish-themed bars, Chinese restaurants and African-American hip-hop music are all big sellers inside the USA. Even the world-famous American hamburger is an ethnic-minority artefact. It is German in origin, despite seemingly being 'as American as apple pie' (and apple pie was in fact brought to America by Dutch and English settlers — long after the Romans introduced apples to northern Europe 2000 years ago).

Geographers studying progressive cultural change in the USA have found that the consequences of diaspora hosting are complex and that relationships change over time. **Assimilation** — when new migrants abandon their own cultural traits in favour of more mainstream ones (typically those of the powerful white Anglo-Saxon 'core' group) — has sometimes taken place. Yet US society is also said to function more as a **melting pot**: rather than some cultures simply assimilating others, an 'American way of life' has evolved over

Figure 5.5 A cultural continuum: differing responses to diversity

Continuum	Response	Description
Progressive acceptance of new diaspora/immigrant cultures	**'Melting pot' (or hybridism)**	Positive view of American culture as organic or hybrid — it adopts and absorbs new migrant values
	Pluralism	EU nations tolerate equal rights for all migrants to practise their religious and cultural beliefs
Cautious acceptance of diaspora/immigrant culture with some controls	**'Citizenship' testing**	UK rules for migrants are becoming stricter in reaction to popular concerns over immigration
	Assimilation	A belief that minority traits should disappear as immigrants adopt host values
	Internet censorship	Preventing citizens from learning about other global viewpoints using online sources, e.g. China
Resistance to increased cultural diversity (right-wing view)	**Religious intolerance**	Notably lower levels of religious freedom for minority groups exist in some places, e.g. Iran
	Closed door to migration	Stopping any immigration altogether for fears of cultural dilution, e.g. Cambodia (the Pol Pot years)

time that incorporates elements of all diaspora cultures present in the USA, be they Italian, Scottish, Jamaican, Native American or Jewish. There is also plenty of evidence for a third phenomenon known as **cultural pluralism**, which describes the way some diaspora groups — such as people of Chinese, Korean or Greek descent — may have been content to maintain distinctive and spatially segregated ethnic identities over time (Figure 5.5).

Activity 5

The multicultural character of the USA as a diaspora host nation is well documented in popular culture. Make a list of all of the American films and television programmes you have seen that portray US ethnic diversity. You could begin by thinking about the characters in the cartoon show *The Simpsons*.

Managing cultural globalisation

Countries are rarely consistent in their response to the costs and benefits of cultural globalisation brought by international migration, trade, media and internet exchanges (Table 5.4). This has been especially true since terrorist attacks on New York (2001), Madrid (2004) and London (2005) triggered fresh debate over the meaning and importance of local identity in a globalising world.

Table 5.4 Benefits and costs of cultural globalisation for selected regions

	Culturally dominant world nations (e.g. UK or Spain)	**Diaspora source nations (e.g. Ireland or Israel)**	**Diaspora host nations (e.g. France, UK or USA)**
Benefits	Successful exporting of cultural traits (e.g. the English language) helps build overseas business opportunities for the nation's companies May positively boost a country's global influence	Enormous potential for tourism exists for nations at the heart of a major world diaspora Global political and cultural influence of small nations like Ireland or Israel is boosted	Immigrant cultures bring new ideas and entrepreneurial energy that can be harnessed, e.g. hybrid music and arts Helps build international bridges and fosters political cooperation
Costs	Accusations of cultural imperialism and neo-colonialism have sometimes made the USA and UK unpopular, even becoming targets for terrorism Can become a 'victim of own success', attracting excessive migrant flows	Too much population is lost overseas, including very talented and creative individuals Culture at home can become stagnant or fossilised — to please the nostalgic expectations of diaspora tourists	Brings too much change to indigenous culture, especially if a 'melting pot' situation of change develops; this can trigger a right-wing reaction and persecution of migrants Post-September 11 2001 there are heightened fears over national security

In the UK, mass migration flows from eastern Europe (following EU enlargement in 2004) have reignited historically recurring concerns that a small country is, to quote ex-prime minister Margaret Thatcher, becoming 'swamped' by immigrants. However, there are many good reasons for the British people to celebrate a multicultural richness that breathes life into new hybrid cultural forms. White British artists have absorbed influences including West Indian reggae and Asian bhangra music and Bollywood cinema. Good economic sense also drives the willingness to mix elements drawn from different cultures in experimental ways. Advertisers enjoy drawing on fresh imagery to sell their products — and immigrant culture can be a rich vein for them to mine.

The consequences of cultural globalisation — in Britain and elsewhere — therefore remain a complex and controversial topic:

- Ought members of diaspora groups be free to fully observe their own religious faiths while living within a host country whose culture and codes differ? Some British Sikhs would like to be cremated in the open air after death, in accordance with their traditions, but UK law will not allow it.
- In France, a key destination for North African Muslim migrants, schools — along with all state institutions — are secular. Debate has focused on freedom for girls of Islamic faith to wear the hijab or headscarf at school. Should French law change to allow this? Should a society-wide ban, as favoured by some politicians, be introduced?
- A cartoon appearing to disrespect the Islamic prophet Mohammed appeared in a Danish newspaper during 2005. Should the newspaper proprietor have been prosecuted for blasphemy?

For questions of culture such as these, there are rarely simple answers.

6 Globalisation and environmental stress

Global economic activity

Economic activity today takes places on an unprecedented global scale. The sheer number of people leading their lives as producers or consumers of commodities has brought a step change in levels of environmental stress. Accelerated cross-border flows of greatly increased volumes of materials and wastes (not to mention many more people travelling in polluting aeroplanes and land vehicles) have massively enlarged humanity's planet-wide carbon footprint. The global problem of heightened carbon dioxide (CO_2) and other greenhouse gas (GHG) emissions is only set to worsen, with China's emissions not expected to peak and decline until around 2040. Scientists now view a 2 °C rise in average world temperature as inevitable — bringing a range of serious impacts for people and places as a result of the enhanced greenhouse effect (Figure 6.1).

Figure 6.1 Predicted impacts of the enhanced greenhouse effect

Temperature rise impacts		When?
+5°C	Nearly half of global animal and plant species under threat due to habitat loss and change	
+4°C	Millions more people exposed to flooding each year — especially in deltas and densely-populated coastal cities	UK Met Office forecasts this for 2060 if emissions are not curbed
+3°C	Greatly increased mortality from droughts and water shortages in semi-arid areas, with millions of 'climate change refugees'	
+2°C	Biodiversity under threat, with 30% of all species at risk, including widespread coral reef bleaching	Forecast for end of twenty-first century if CO_2 level can be kept at 450 parts per million
+1°C	Increasing risk of wildfires and evidence of plant and animal species migrating due to habitat changes	

Modern globalisation is not directly responsible for starting this process. It was western Europe's early industrial revolution after the 1750s that lit the fuse for global atmospheric change (according to ice-core evidence). However, today's global networks of production and consumption, relying as they do on the perpetual movement of container ships, aeroplanes and megatrucks filled with mass-produced consumer goods, have accelerated GHG emissions, resulting in the 'hockey-stick' temperature-rise scenario with which we are all too familiar (Figure 6.2). It tells tales of excessive food miles clocked up by agricultural produce and cheap no-frills airlines flying increased numbers of well-to-do pleasure-seekers around the world (in the EU alone, carbon emissions from air flight doubled between 1990 and 2006).

Figure 6.2 'Hockey-stick' trends for CO_2 and average world temperature

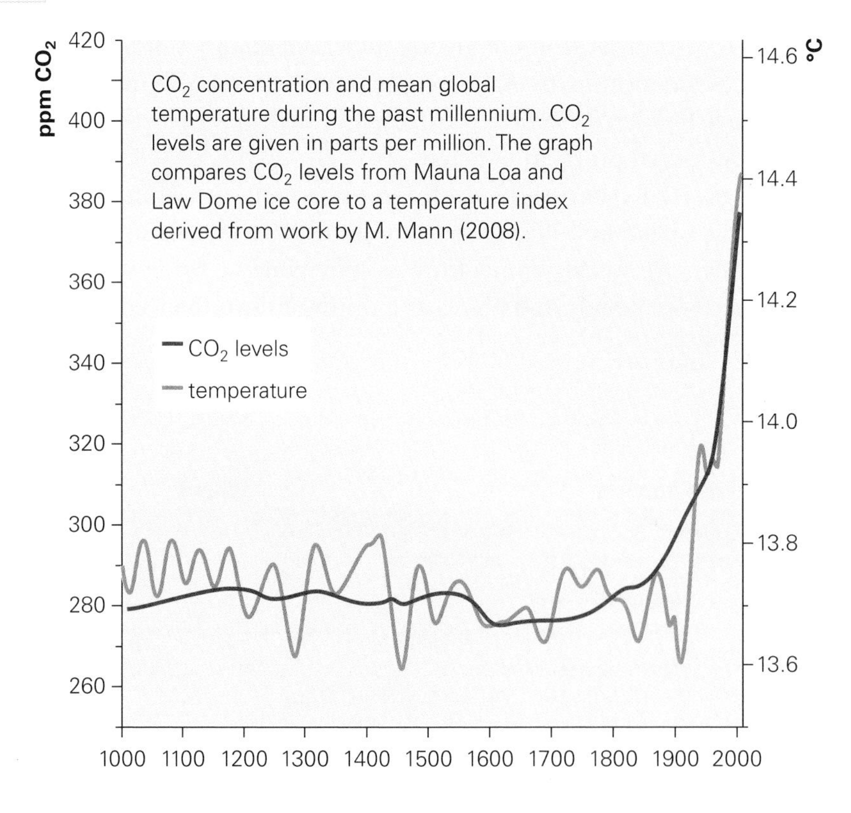

Few places on Earth do not bear some local sign of stress or damage as a result of globalised activities. The most notorious events have been well reported, often becoming the subject of protracted legal action or justice campaigning, or covered by documentary film-making (Table 6.1).

Table 6.1 Selected pollution events and their global connections

	Pollution event	Global connections
1978	The ***Amoco Cadiz*** tanker oil spill affected 72 km of French beaches. Sunk by a storm, the tanker was responsible for US$250 million damage to fishing and tourism and killed 20 000 birds.	1.6 million barrels of crude oil were spilled by this supertanker en route from the Persian Gulf to Rotterdam in the Netherlands. The English Channel is a busy shipping lane and its coastline is an oil-spill hotspot.
1984	Over 7000 residents of the Indian city of **Bhopal** died when a deadly plume of methyl isocyanate (MIC), a by-product of the pesticide manufacturing process, was accidentally released from a factory. Thousands more have died since due to long-term effects.	The pesticide plant in Bhopal was owned by the American TNC Union Carbide (a business that was subsequently acquired by the even larger chemical giant Dow Chemicals). Low labour costs attracted Union Carbide to Bhopal as part of its spatial division of labour.
1993	Reports of serious degradation of **Nigeria's Ogoniland** due to oil spillages first began to emerge. Indigenous writer Ken Saro-Wiwa led the protests that gained media attention; he was executed by Nigeria's government in 1995, causing an international outcry.	Transnational oil firms — including Royal Dutch Shell, BP, Total (France), Italy's Agip, and ExxonMobil and Chevron from the USA, have frequently been accused by NGOs such as Amnesty of bringing great environmental damage to Nigeria and other countries.
2007	Cargo container ship ***MSC Napoli*** beached close to Branscombe on the English 'Jurassic Coast' (a UNESCO World Heritage site). Motorcycles and car parts were washed ashore as well as packets of oil-covered biscuits, which caused sea birds to fall ill after ingesting the oil.	*MSC Napoli* was built in South Korea by Samsung Heavy Industries and owned by a company based in British Virgin Islands. The ship was carrying commercial and personal containerised goods from Belgium to Portugal through the English Channel when disaster struck.

Activity 1

Just how much GHG pollution is a country like the UK really responsible for? Government figures reflect only 'domestic' emissions (gases emitted within the UK's borders), putting the UK's contribution at 2% of the global total. Even this is not a figure to be proud of, given that the UK is home to just 0.1% of world population. However, this estimate does not include emissions resulting from UK firms' global activities. Might a more honest figure be 4% or higher, as critics suggest? Research this global issue online. See these websites:

www.christianaid.org.uk/Images/coming-clean-uk-carbon-footprint.pdf

http://news.bbc.co.uk/1/hi/sci/tech/8283909.stm

The global relocation of polluting industries

Wherever possible, TNCs seek low-cost locations for their manufacturing and refining operations, with cheap labour, as we saw in Chapter 4, often a key consideration. However, another attractive location factor can be weak environmental controls. In high-income nations, bodies such as the UK Environment Agency are well funded to carry out their brief of closely monitoring industrial operations. In comparison, highly polluted sites such as Nigeria's Niger delta — where Amnesty International estimates that nearly 7000 spills occurred during the 1980s and 1990s, harming both people and one of the world's ten most important coastal habitat zones — clearly lack much-needed environmental control. A selective snapshot of the world taken in 2009 reveals the everyday nature of polluting activities associated with the global economy:

- In China's rural Hunan province, many people show signs of poisoning by a lead-emitting manganese smelter (manganese is used to strengthen steel, one of China's major global exports).
- In Accra, Ghana, entire families undertake the dangerous work of breaking up old computer monitors imported from Europe and melting down circuit boards to extract the metals — leaving large amounts of toxic waste.
- In Ivory Coast, tens of thousands have suffered ill health after toxic waste alleged to produce hydrogen sulphide was dumped by a ship in the employ of Trafigura, a European TNC.

Sometimes, justice is sought and won. Ivorians finally reached a £28 million settlement with Trafigura late in 2009. However, the company continues to deny direct accountability, insisting instead that a third-party waste-disposal contractor in Trafigura's employ made the decision to visit Ivory Coast independently. As we saw in Chapter 4, outsourcing has sometimes allowed TNCs to wash their hands of unethical actions carried out in their name by the companies they employ.

For similar reasons, the people of Bhopal have yet to receive adequate compensation for their exposure to one of the worst industrial disasters on record (see Chapter 4, and Table 6.1, p. 73). Acquisitions such as that of Union Carbide by Dow Chemicals are a routine way for successful TNCs to build market share, power and influence. They also provide the perfect pretext for legal evasion: Dow claims no direct responsibility for Union Carbide's past actions in India. Not only has Dow resisted making large payouts to the

Bhopal petitioners, it has also failed to clean up the site (although renewed pressure from the US Congress for the company to be held accountable may yet change the outcome of this case).

Activity 2

A detailed Amnesty report on oil pollution in Nigeria can be downloaded at: **www.amnesty.org/en/library/asset/AFR44/017/2009/en/e2415061-da5c-44f8-a73c-a7a4766ee21d/afr440172009en.pdf**.

National Geographic reported on Nigeria in 2007 at: **http://ngm.nationalgeographic.com/2007/02/nigerian-oil/oneill-text.html**.

Use these readings as the basis for either a PowerPoint presentation or an extended essay on the subject.

Transboundary pollution

A transboundary pollution event is one that has damaging effects for more than one country. Climate change is perhaps the ultimate example of this. Degradation of water quality in transboundary aquifers, of which there are 273 in the world, is another growing concern. For instance, the Guaraní aquifer underlies 1.2 million square kilometres of land shared by Brazil, Argentina, Paraguay and Uruguay. This vital water store is under pressure from many globalised activities such as cattle rearing for international markets.

Acid rain

Acid rain (precipitation with a pH value below the naturally occurring level of 5.6) is a type of transboundary pollution commonly caused by anthropogenic sulphur emissions, which have risen worldwide from 5 to 180 million metric tonnes since 1860. In heavily industrialised and polluted regions, acidity as high as pH4 may be expected: Scotland once recorded an extreme value of 2.4 prior to deindustrialisation. Impacts of acid rain for aquatic ecosystems can be highly damaging, especially where local geology lacks any alkaline buffering capacity. Freshwater lakes underlain by silicate granite or quartzite rocks are worst affected, with severe acidification resulting in impaired reproduction and tissue damage for fish, leading to biodiversity reduction. Species such as trout, roach and perch have vanished altogether from some northern European lakes on account of acid rain.

In recent decades, taller industrial chimney stacks have been built; these reduce the intensity of local impacts but instead allow acid rain to spread pollution over a wider geographical area, albeit at lower intensity. Local pollution problems can become internationalised as a result. For instance, during the winter of 1952, 4000 bronchitis-related smog deaths were recorded in the then heavily polluted industrial city of London. The construction of taller stacks was one of many mitigation measures subsequently adopted. However, British pollution now impacts on northern near-neighbour Sweden as a result: upper-air currents carry sulphur emissions northeastwards, following the jetstream (Figure 6.3). Similar diffusion patterns for pollution have been recorded in North America since the 1950s, when only two stacks found there were higher than 180 m. Now hundreds of stacks exceed 180 m, including the copper-nickel superstack smelter in Sudbury, Ontario. Owned by the TNC Inco, this monster chimney is 380 m high. Before measures were introduced in the 1990s, it fed 1% of all global sulphur emissions into high-level air currents.

Figure 6.3 Acid rain over Scandinavia

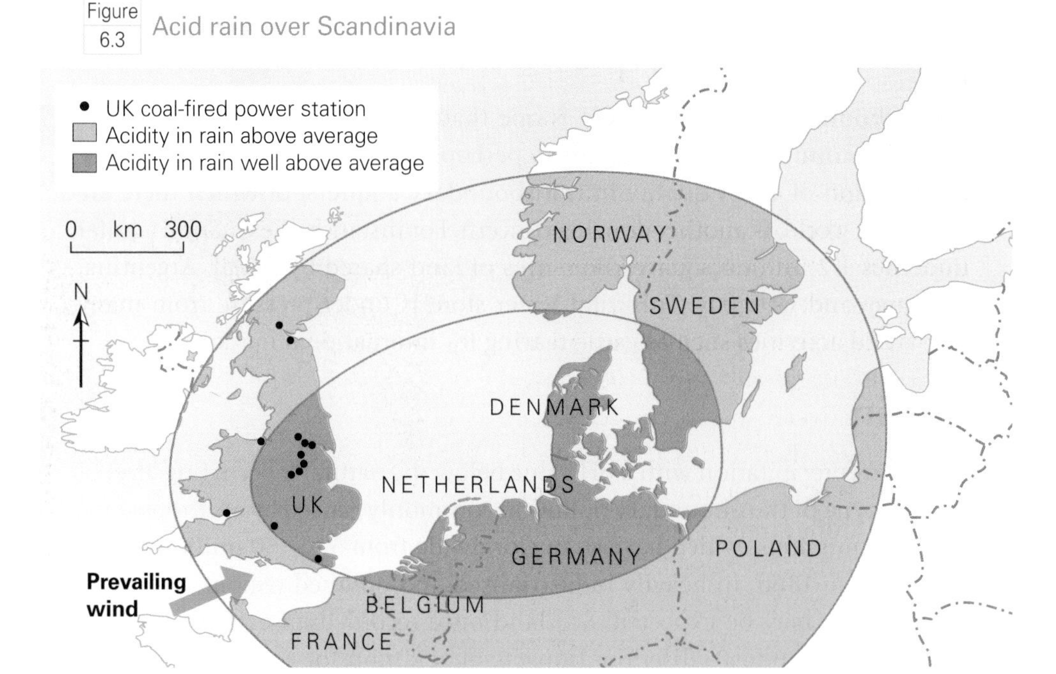

Oil and the environment

World trade in oil has been a great economic success story for nations controlling its flow production, such as the United Arab Emirates (UAE) and Saudi Arabia. It is also one of the most environmentally damaging global activities. First, consider the role of fossil fuels in creating an enhanced greenhouse effect. Second, remember that terrestrial oil exploitation has been a major cause of land degradation and air pollution in regions such as the Niger delta due to spillages and flaring (the practice of burning off gas released in oilfields, and a major cause of transboundary acid rain — see Figure 6.4). Third, examine how bulk oil movements have brought scores of devastating transboundary pollution events to those territories flanking shipping lanes since supertanker technology developed after the 1950s.

The coastal margins of both France and the UK were severely affected by the 119 000 tonnes of oil released from the *Torrey Canyon* supertanker in 1967, after it struck a reef in the English Channel en route from Kuwait to Milford Haven in the UK. This was the first major oil spill to make world headlines. Fifteen thousand seabirds were killed, and 80 km of UK beaches and 120 km of French coastline were contaminated. This remains the UK's worst ever environmental disaster to date.

Tackling transboundary pollution

Reaction to such events by world governments and civil society often takes the form of outrage and calls for greater regulation. One area of success has been the voluntary adoption of the United Nations Convention on the Law of the Sea (UNCLOS) by 156 signatory states. UNCLOS legislation makes it illegal for ships that have recently delivered oil to use seawater to wash out their tanks (flushing of tanks has been a significant cause of oil pollution along major shipping lanes). Seafaring vessels are also required to meet improved standards. One particular achievement has been progressive retirement of the worst-offending single-hulled oil tankers that were too easily damaged in the past. The last major single-hulled disaster occurred in 2002, when the *Prestige* supertanker sank off the Galician coast, leading to the largest environmental disaster in Spain's history.

Figure 6.4 Environmental pollution is caused by burning off gas from oil production in the Niger delta

Friedrich Stark/Alamy

Global phase-out of older tankers is well under way. Yet even when taken out of service, these leviathans leave behind a legacy of ecological and human harm. Vessels are routinely sent to India, Pakistan and Bangladesh, where ship-breaking is carried out at low cost by poorly paid labourers. Toxic materials leak from filthy hulls, polluting the coastal waters where the ship-breaking yards are located. Environmental group Greenpeace has spoken out against this practice and is campaigning for improved governance.

Agribusiness and biodiversity loss

Though rarely generating the same dramatic headlines that chemical spills attract, the transformation of 40% of Earth's terrestrial surface into productive agricultural land brings ecological change to environments everywhere. Food chains and nutrient cycles have been modified across the globe, often bringing habitat loss and biodiversity decline on a continental scale.

Much of this land is now in the hands of major global **agribusiness** (a blanket term covering various types of agro-industry including food, seed and fertiliser production, as well as distribution and wholesale). The larger agribusinesses are TNCs such as Del Monte. Their extensive production networks are able to source globally food that will be available to European and North American shoppers on supermarket shelves. The largest firms, when measured in terms of their controlling land interests and revenues, exhibit enormous size and power (Table 6.2).

Table 6.2 Some large agribusinesses, their activities and turnover

Name	Headquarters	2008 revenue (US$ m)	Activities
Kraft Foods	Illinois, USA	42 200	Food processing
Unilever	Rotterdam, Netherlands	40 500	Conglomerate
Nestlé	Vevey, Switzerland	109 900	Food processing
Carrefour	Levallois-Perret, France	89 000	Food retail
Cargill	Minnesota, USA	120 000	Food production and technology

The impacts of these firms penetrate deeply into many of the world's poorest regions, such as east Africa and southern Asia (Figure 6.5). A variety of activities, including the intensive production of cash crops, cattle-ranching and aquaculture, bring damaging environmental effects:

- **Monoculture and habitat loss** With 40% of the world's land surface now used for farming, habitat loss has been experienced by many of the planet's

Figure 6.5 Impacts of agriculture, food and drink TNCs operating in parts of Asia

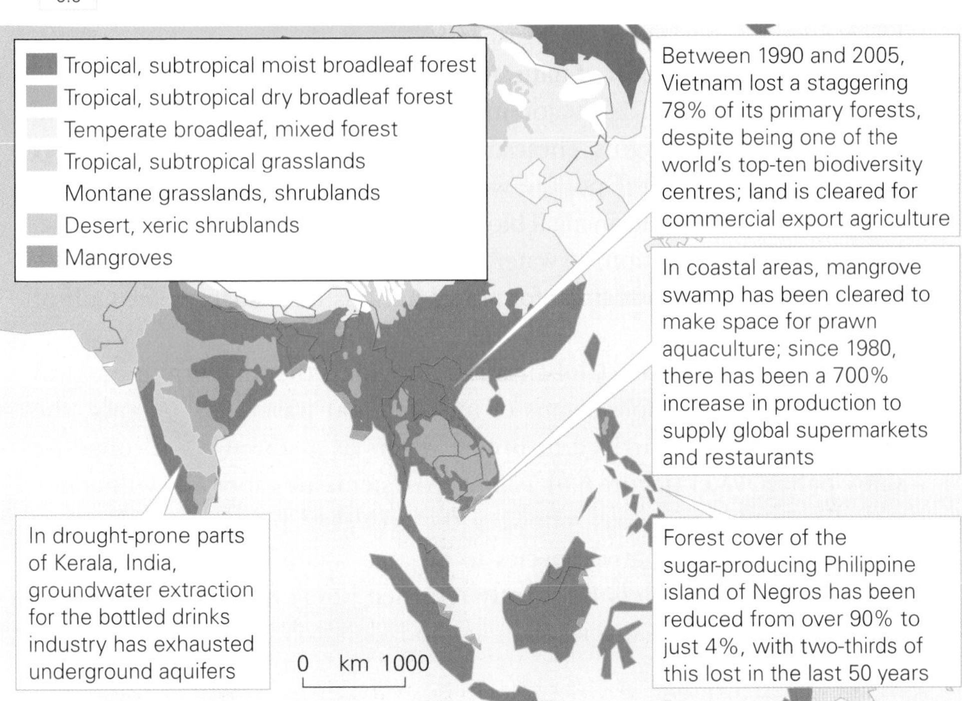

1.4 million identified species. As a result, thousands of species are now in danger of extinction. The largest agribusinesses have promoted wheat, maize, rice or potato monoculture (both shaping and reflecting the homogenisation of diet that is an aspect of cultural globalisation). These four crops now account for 60% of plant-derived calories in human diets worldwide. Global genetic diversity is vital for lasting ecological sustainability but is clearly threatened by this.

- **Water store depletion** The worst effects occur when intensive crop farming is introduced to areas where only limited water supplies are naturally available. Groundwater abstraction by Coca-Cola may have contributed to water scarcity in parts of Kerala in southern India (see Figure 6.5). The term 'virtual water' is used to describe the water 'embedded' in the production of food or goods for global markets. Each UK citizen consumes 4600 litres of virtual water every day, according to one estimate, giving the country as a whole a large water footprint.
- **Loss of natural protection** Forests provide vital ecosystem benefits for people and places. Rising incidence of flooding in the Ganges delta can be linked to timber removal, with much of the wood feeding global hardwood demand. Mangrove forest similarly offers protection to tropical coastal

margins — in this case, against storm surges. However, insatiable world consumer demand for tiger prawns leads directly to mangrove clearance in places like Indonesia and Madagascar. Space is created for prawn aquaculture ponds but at the cost of natural tsunami protection.

- **Eutrophication** About 20 major 'marine dead zones' lie scattered around the world's coastal margins. These are sites where intense inputs of fertilisers from agricultural runoff pollute waterways and overstimulate ecosystem productivity. This results in algal bloom growth and its subsequent collapse, leading to deoxygenation of water and species death. Some of the worst affected areas are hub regions for global agribusiness, such as the Gulf of Mexico.
- **Invasive alien species** Globalisation provides the modern ecological hitchhiker with multiple means of international travel. For instance, the Chinese mitten crab arrived in British waters as a stowaway in container ships' ballast water (Figure 6.6). Island ecosystems are especially vulnerable to indigenous species loss following the arrival of invaders. In the past, introduction of non-native species to an area was often a deliberate act: British colonial explorers frequently returned home burdened with seeds and cuttings of exotic species such as rhododendron, giant hogweed and, of course, the potato.

Figure 6.6 Examples of marine alien invasions

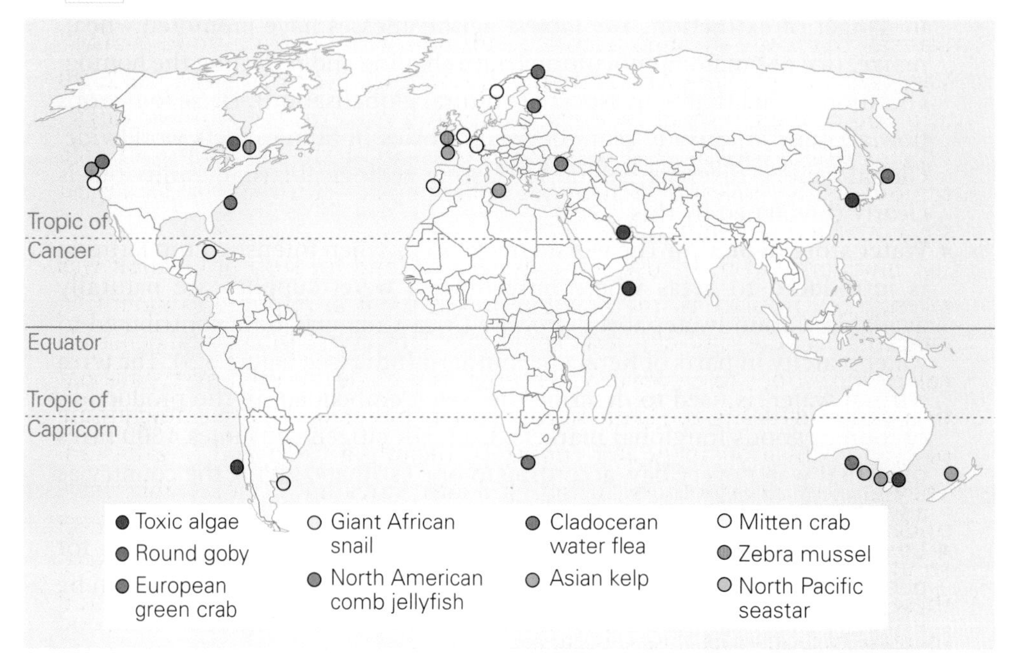

- **Genetically modified organisms** GMOs were first grown commercially in the mid-1990s in developed countries. TNCs now cultivate them in developing countries such as Argentina, Brazil, India and South Africa. GM crops, engineered for greater herbicide tolerance or insect resistance, can be expected to reduce biodiversity wherever they are introduced.

Rising affluence and 'land grabs'

Integral to the bigger picture of globalisation and environmental stress are population changes. These encompass: growing numbers of longer-lived people with access to improved healthcare and diets; rising numbers being lifted out of subsistence poverty (even if the gap is widening between Earth's richest and poorest); and increased individual aspirations (as consumerism sweeps the planet).

These points are well illustrated by an analysis of the changes currently underway in China. As we learned in Chapter 4, 400 million Chinese have been lifted out of absolute poverty since the 1980s. Although most may still receive low incomes by Western standards, significant numbers have reached a level where the dietary shift from basic vegetarianism to include dairy and meat consumption becomes possible. This markedly increases agricultural land demands. Raising cattle to feed one person meat requires inputs of grain and land that might otherwise feed ten (due to inefficient transfer of energy through trophic levels within agricultural food chains). Vast tracts of pristine Amazonian rainforest were cleared between 2000 and 2006 to make space for soya production used to feed Chinese cattle. Brazil's Mato Grosso state experienced particularly rapid rates of tree removal (state governor Blairo Maggi happens to run the world's largest soya production company).

A moratorium on the use of newly cleared land for soya production was recently put in place by the Brazilian government in response to mounting concerns over the scale of forest devastation. However, China's food demands can only continue to grow as greater progress is made towards poverty alleviation. Mindful of this, the Chinese government has embarked on a programme of land acquisition in poorer countries, including Cuba and Kazakhstan. They are not alone; others, including South Korea and Saudi Arabia, have undertaken similar ventures in response to heightened concerns over food security (Figure 6.7). Described as the **land grab** phenomenon by newspapers, these neocolonial enterprises are certain to accelerate biodiversity loss and increase carbon emissions (Figure 6.8).

Figure 6.7 The global food crisis, 2008

Population growth
World population grew from 2.6 billion in 1950 to 6.7 billion in 2008. It will reach 9.3 billion in 2050

Rising affluence
The middle classes in growing economies like Brazil, Russia, India and China want more food than before, especially meat and dairy

Food crisis, 2008

Energy policy
Some crops are being used to produce biofuels instead of food

Climate change
Extreme weather devastated crops in Myanmar and parts of Africa in 2008. Some events may be due to climate change, others are 'normal' climatic events

Figure 6.8 World land grabs: buying foreign fields in 2009

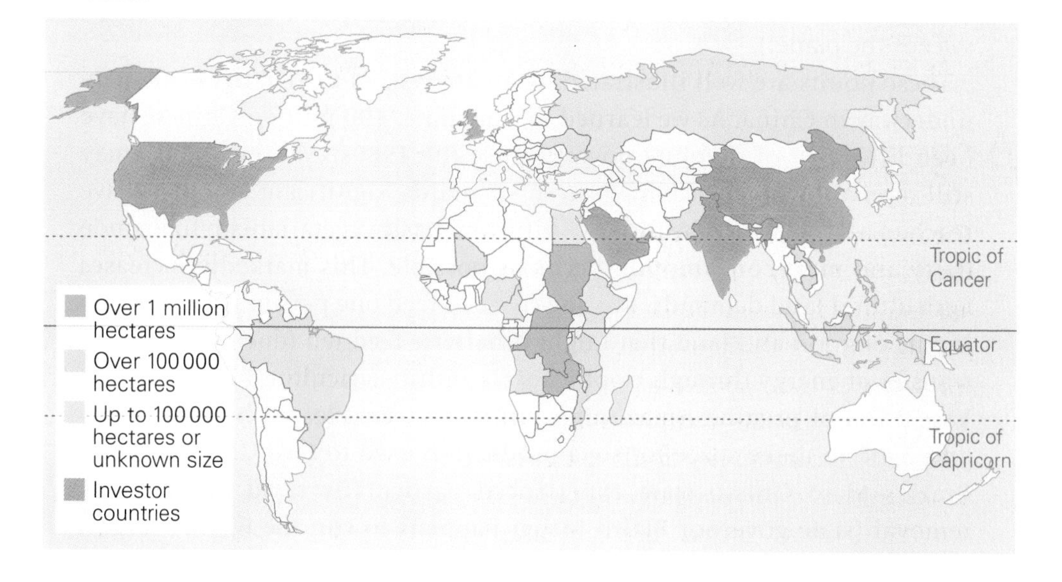

Activity 3

Can vegetarianism save the planet? The UN Food and Agriculture Organization (FAO) study *Livestock's Long Shadow* suggests animal husbandry is responsible for a staggering 18% of all anthropogenic greenhouse gas emissions, in addition to placing pressure on water supplies in areas of scarcity. Worldwide, the number of farm animals has grown fivefold since 1950 to keep pace with humanity's growing dairy and meat demands. Taking affirmative action, the Belgian city of Ghent declared Thursday its weekly 'meat-free' day in 2008. Use these facts as a starting point for investigating the claim that 'rising meat and dairy demand is Earth's greatest threat'.

The rising affluence of the world's population has also led to an increase in:

- **Tourist flows** Air flight costs have fallen over time, while affluence has increased for many, making travel to distant places more affordable. Expansion of the 'pleasure periphery' (remote regions of the world, often possessing wilderness qualities) places stresses on previously undisturbed fragile and unique environments. For instance, visitor number growth from 200 000 to over 500 000 per year has led to Peru's Machu Picchu World Heritage Site being placed on the UN's 'endangered' list.
- **Recycled waste flows** UK ships deliver 40 million tonnes of refuse to China for recycling each year (60% of the UK's plastic waste and 50% of its steel waste is exported for reprocessing overseas). The need to export waste arises because of mounting legislation deterring local councils from adding to existing landfill, plus a general lack of recycling facilities in the UK where, in common with other manufacturing operations, running costs are high. As a result, garbage has experienced its very own global shift.

Are we living sustainably?

The answer unquestionably is no (Figure 6.9). Humankind's collective ecological footprint has exceeded planet Earth's biocapacity. This means that current levels of resource use and waste and pollution generation are simply unsustainable in the long term.

Figure 6.9 A model of sustainable development

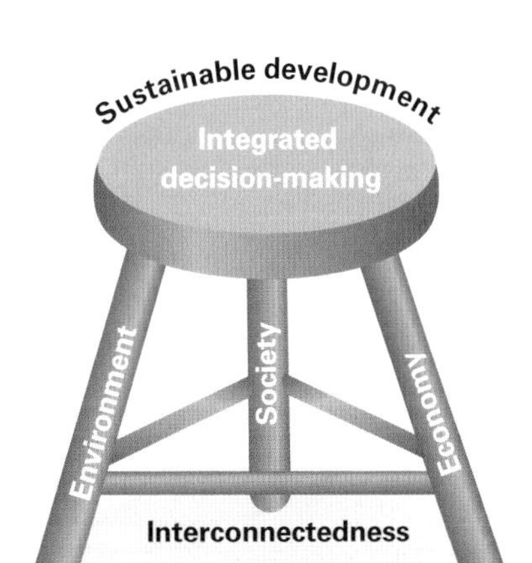

The global community needs to clean up its act urgently. Even apparently harmless activities such as internet use can carry a high environmental price tag, with one recent report suggesting that every query posed to Google results in 7 g of carbon emissions. The worst predicted impacts of climate change threaten society on a planetary scale, while most localities increasingly bear signs of stress brought by global commodity or waste flows. For instance:

- Plastic toy ducks have been washed ashore on once-pristine Alaskan beaches long after a flotilla of toys was set adrift following a container ship accident in the Pacific Ocean in 1992 (Figure 6.10).

- The Pacific garbage patches are two vast stretches of water, far from any land, where plastic and other debris washed out to sea from cities all over the world have been collected and bound in place by circulatory ocean currents (Figure 6.11).

Figure 6.10 The Pacific orbital path followed by a spilled container of plastic ducks

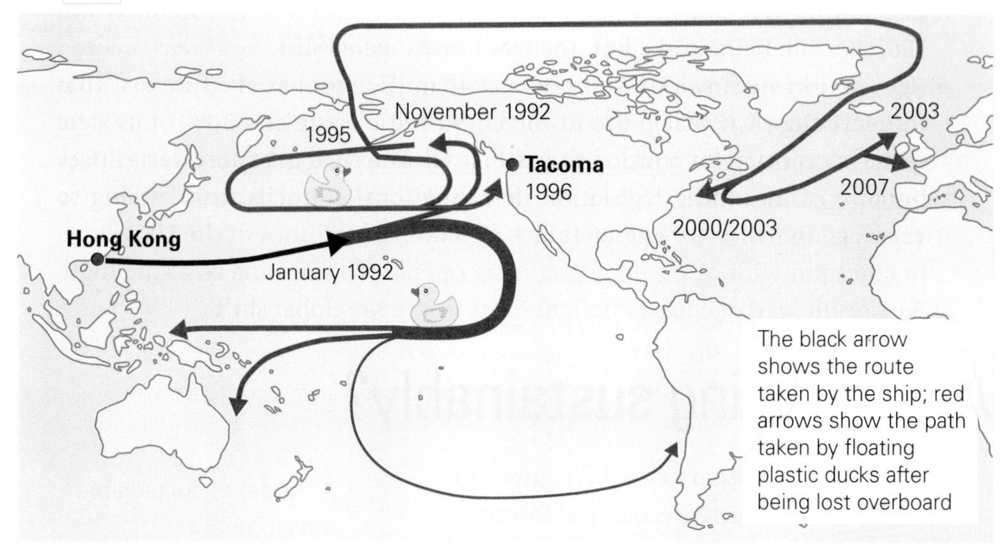

Figure 6.11 Pacific garbage patches

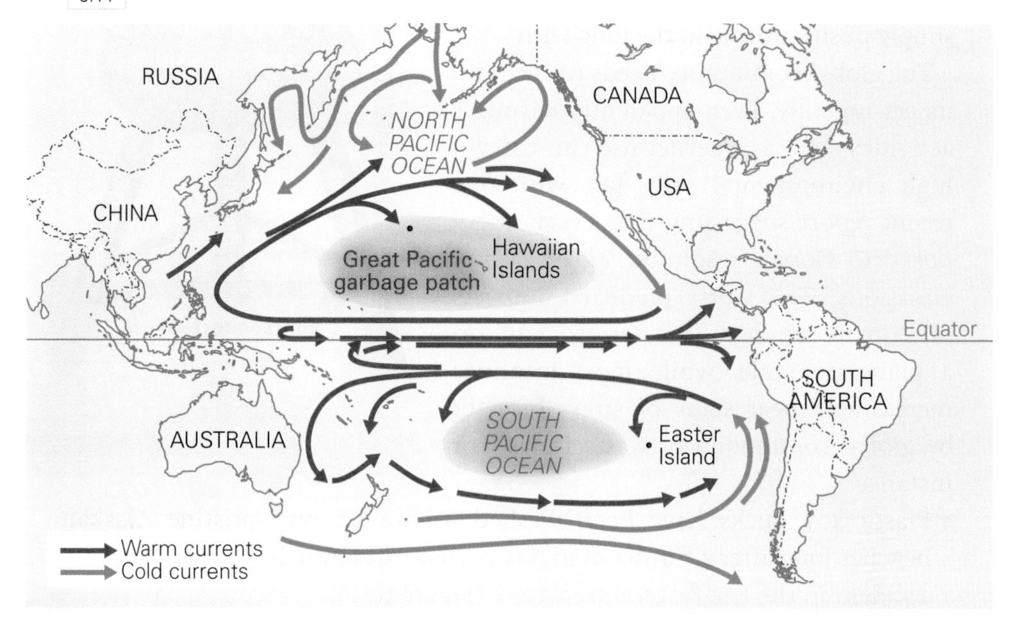

Globally polluting output of manufactured goods and services will only increase in future years as more wealth is created, notably in the emerging economies. Meanwhile, our flagship resource conservation strategy — recycling — does not prevent new environmental damage; it merely slows the rate at which fresh resources are utilised. (Remember that the recycling process itself demands the carbon-emitting operation both of transport and energy-hungry machinery.)

Are there any grounds left for optimism? Economic optimists believe markets will find a solution to the sustainability problem. It is argued that given sufficient financial incentive, the business community will fund the research necessary for much needed breakthroughs in cheap, clean energy production to be made, such as improved solar power technology or effective carbon capture and storage capability for power plants.

Activity 4

Study Figures 6.10 and 6.11. Explain how human and physical factors have contributed to the pollution of wilderness and remote places on a local and global scale.

7 Political outcomes of globalisation

Global groupings: the emergence of trade blocs

The United Nations recognises 193 countries in the world; all are networked together to varying degrees by diverse global financial, commodity, population and information flows. At first take, this might suggest a potentially chaotic state of twenty-first century affairs — as suggested by the phrase 'all that is solid melts into air'. In fact, however, a degree of structure has been brought to the proceedings. Like the lunchtime playground, an apparently busy yet disorderly crowd is, on closer inspection, found to be composed of groups and cliques, much of whose energy is focused on their own internal affairs.

Figure 7.1 Selected regional trade bloc groupings

So too do we find that strongly regionalised activities are a major part of the pattern of global interactions (Figure 7.1). This is because most countries, whether rich or poor, choose to join broad economic groupings known as **trade blocs** (Figure 7.2).

Transactions between group members are conducted on preferential terms, with reduced tariffs placed on imported and exported parts and goods, or with tariffs eliminated altogether. Exchanges still take place between group members and non-group members — quite notably evident in the case of an enormous surplus of Chinese goods imported into the EU and North American Free Trade Association (NAFTA) states — but with taxes and import quota restrictions usually remaining in place. The fact that imports from China are so attractively priced, even with tariffs in place, is a reminder of just how low Chinese production costs are.

There are now over 30 major trade blocs and agreements in existence, all exhibiting varying degrees of market liberalisation and customs harmonisation (Figure 7.3). Any decision by states to participate openly in free trade is taken knowing that the promise of easier international sales for firms sits alongside an increased risk of foreign goods flooding home markets.

Figure 7.2 African trade blocs, with some of the least developed countries as members

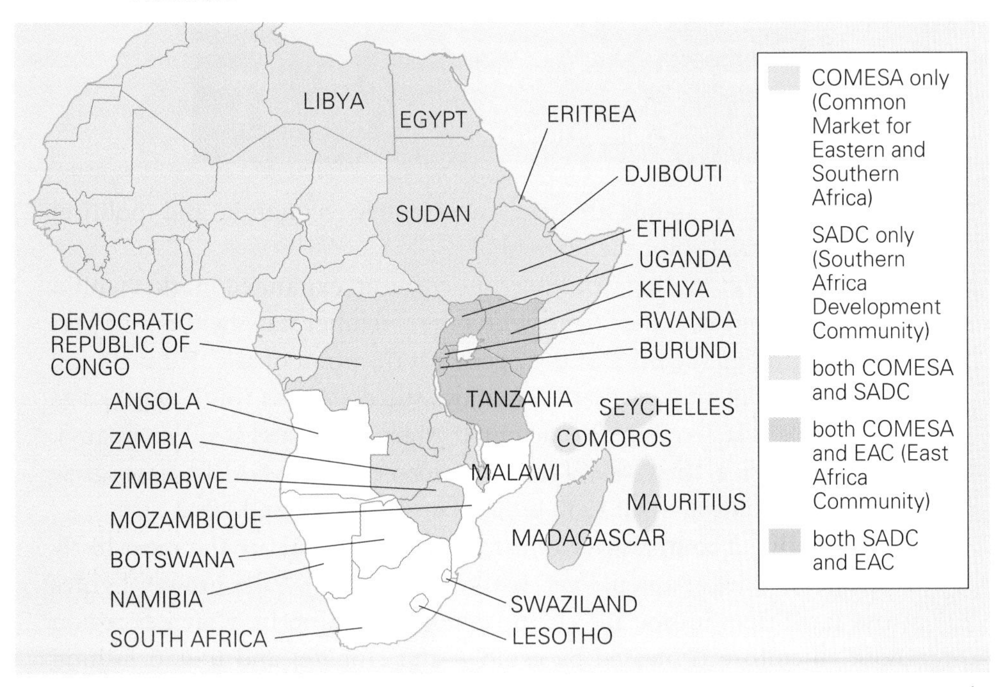

However, the overarching logic of the agreement — enshrined in Ricardo's theory of comparative advantage — dictates that all member-state companies and citizens should, on balance, find themselves net beneficiaries of the new economic order. Thus a degree of economic autonomy is willingly ceded by all parties. Memory of the Great Depression of the 1930s (when protectionism was favoured above free trade, and many countries placed punitive taxes on foreign imports to help prop up employment in home industries, even if these were outmoded or uncompetitive) is often cited as a ringing endorsement of the merits of free trade.

Figure 7.3 'Layering of complexity': the different types of multi-governmental trade bloc

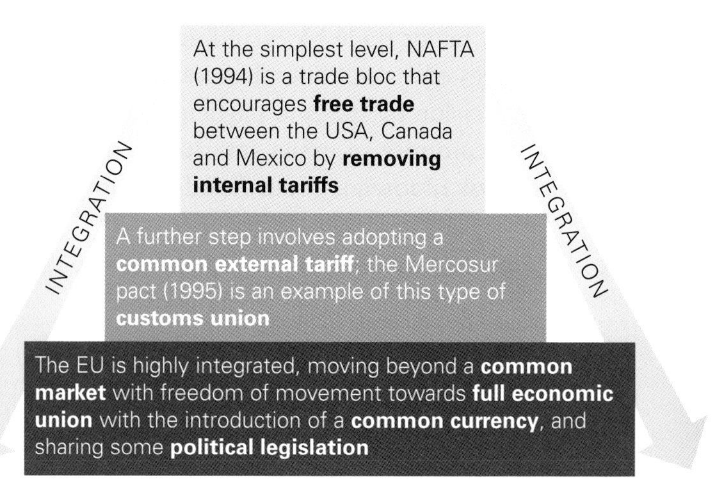

A sovereign state stands to gain considerable economic and political advantage from trade bloc membership:

- Successful market-leading firms may anticipate an **expanded market** within a trade bloc. They can trade freely with other member states without import tariffs pushing up the price of their goods at the point of sale. For instance, Brazil's energy giant Petrobras now enjoys success across South America as part of Mercosur (having also acquired Argentina's largest oil company). In North America, the value of trade between the three NAFTA states grew from US$306 billion to US$930 billion between 1993 and 2007.
- Firms boasting a **comparative advantage** of some sort are the most likely to prosper as part of an enlarged market (Figure 7.4). Sales growth figures across tariff-free Europe for champagne (originating in a French region whose viticulture is blessed by geology and climate) and Italian fashion (which enjoys the ready availability of high-quality leather) are often

Figure 7.4 Comparative advantage: the development of trade specialisation within a trade bloc

Pre-trade bloc

All four countries produce their own goods and services across all four sectors (vehicles, textiles, financial services and wine); trade between one another at first is non-existent. As a result, output is limited in all cases and product costs are high for consumers. In some cases the goods and services may well be of poor quality, possibly due to inefficiencies resulting from a lack of high-calibre physical or human resources.

Trade bloc (with customs union)

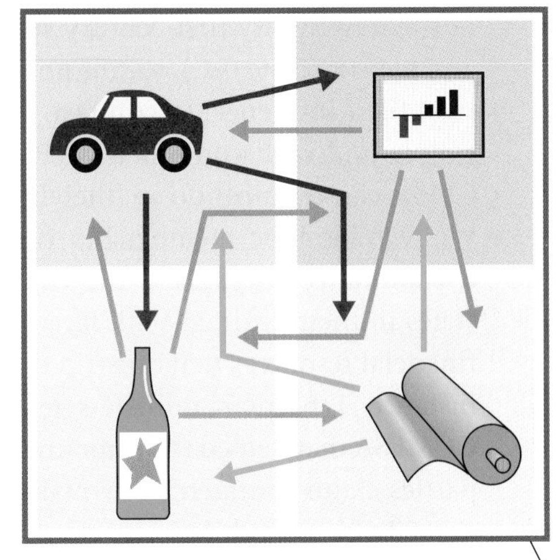

A trade agreement is now in place. National borders are rendered permeable for trade flows through removal of tariffs. Each country discovers it has a **competitive advantage** in one of the four sectors; these firms emerge as leaders in an enlarged market. Scales of production increase and costs of goods and services for consumers fall. A common external tariff may protect these 'winners' from overseas imports.

quoted as examples of this logic. In contrast, poor-quality goods and service providers are less likely to survive as independent businesses.

- As sales rise, successful firms can exploit an **economy of scale** in production. When sales were restricted to small national markets, businesses might have performed at under-capacity. The costs of lighting, heating, staff salaries and so forth would have been met by profits skimmed off a relatively limited volume of sales, thereby driving up the product unit price for consumers. However, when the firm's output rises within a trade bloc, largely unchanged fixed costs (or 'overheads') are paid for out of revenue drawn from an increased volume of sales. This allows firms to price their products more competitively. Consumers living within the trade bloc area also become beneficiaries as they start to enjoy lower prices.
- EU states have one further important advantage, which is a strengthened **collective bargaining** position within international decision-making

organisations. The EU occupies eight seats at G20 meetings. In addition, three of just eight IMF directors who represent single states are from EU countries. In both cases, this gives the EU strong leverage for brokering global economic agreements likely to benefit European firms and states. The size and strength of the two twenty-first-century single-state superpower economies, China and the USA ('the G2'), means no single European country acting alone can rival their influence. In contrast, the EU, representing nearly half a billion people and with combined GDP greater than the USA and twice that of China, can certainly do so (thereby making 'the G3').

- While trade bloc membership can threaten individual state sovereignty, it simultaneously fosters the **capability-building** of poorer and weaker states in their dealings with large TNCs, the most powerful of which have financial resources that outstrip the wealth of many nations (Table 7.1). For instance, Ethiopian ministers received training through membership of COMESA that helped them mount a legal challenge against Starbucks (both parties claim commercial ownership of the coffee brand Sidamo, which is named after an Ethiopian region).

Table 7.1 How the wealth of TNCs compares with that of the world's countries

World's top five TNCs (2008)	
Name	**Revenue (US$ bn)**
Exxon Mobil	477
Royal Dutch Shell	458
Wal-Mart	404
BP	367
General Motors	148

2008 GDP of selected nations		
Name	**GDP (US$ bn)**	**Rank**
Thailand	261	34
Nigeria	212	38
Pakistan	168	46
Bangladesh	79	60
Zimbabwe	3	144

The trend towards trade bloc membership has accelerated since the 1980s. Technology has been a key enabling factor, with improved transport and communications helping firms build wider spatial networks. Budget airlines and a continental high-speed rail network have been especially important for the evolution of the EU. Europe's leisure and holiday providers have prospered alongside real-estate developers; Britons can shop effortlessly for second homes in the Mediterranean, thanks to easyJet, Eurostar and the internet. Elsewhere in the world, highway construction has played a vital role in trade bloc consolidation, notably so in South America.

Activity 1

Study Figure 7.4.

(a) To what extent is this model applicable to the European Union? Which countries may have a comparative advantage for the sectors shown?

(b) Suggest reasons why some indigenous businesses might disappear following trade bloc formation.

Activity 2

There are many trade blocs and trading agreements to research. This is an ideal topic for seminar work: you could prepare a PowerPoint presentation on the history, structure, functioning and consequences of different examples. Possible suggestions include the Andean Community, APEC (Asia–Pacific Economic Cooperation), Cairns Group, EEA (European Economic Area), GCC (Gulf Cooperation Council), the Arab Maghreb Union or SADC (Southern African Development Community).

Geographical consequences of trade bloc formation

Trade bloc growth brings territorial reorganisation as firms expand, merge and create new spatial divisions of labour. Wages in parts of eastern Europe or Mexico are around one-tenth of those in western Europe and the USA respectively, inevitably encouraging relocation (or expansion through new investment) of certain types of low-skill operation. For example, chocolate manufacturers increasingly favour production within Europe's low-wage periphery. Other common effects include the following:

- **Deindustrialisation** Spatial shifts in production create unemployment in places where jobs are lost. The US state of Arizona has seen several of its manufacturers relocate to Mexico, while many more have outsourced the production of goods to Mexican firms. Rationalisation is another response that threatens jobs, recently exemplified by Unilever's 2008 decision to close its historic Dijon mustard factory in central France and increase output in other sites.
- **Mergers and acquisitions** Increased numbers of corporate takeovers bring change to the economic landscape of trade blocs. The Mini, once a leading brand in the UK car industry, is now built by Germany's BMW company. Over time, large businesses such as Unilever or the accountancy firm KPMG have grown to become complex geographical entities within Europe.

- **Post-Fordist production** Some firms have shifted from mass manufacturing into niche or 'premium' markets — alternatively described as a shift from Fordist to post-Fordist production. Instead of attempting to build mass sales in their home market, EU firms concentrate instead on growing a profitable specialist market spread across all 27 member states. The handmade equestrian gear of Cornish firm Mattes is one example; the profusion of expensive low-output regional meats, cheeses, ales and wines on sale in specialist food shops throughout Europe is another.

In the exceptional case of the EU, member states enjoy even greater economic benefits. Major shared policies are in place for, among others, the funding of agriculture, energy and regional development. As a result, nations may:

- participate in the EU carbon pricing scheme (a cap-and-trade system delivering some states surplus carbon credits that they can sell)
- receive a range of agricultural subsidies and support through the Common Agricultural Policy (CAP)
- apply for EU structural funds to develop infrastructure and regional capability (including grants for ICT, or business training)
- easily recruit migrant labour, both skilled and unskilled, from other EU states under the Schengen Agreement

Activity 3

(a) Conduct an analysis of the geography of one or more of Europe's largest firms. Possible case studies include Unilever, Nestlé, BMW or Tesco. Identify their main production sites within Europe, and their major markets.

(b) Explain how the establishment of the EU has encouraged growth for the firm(s) you have studied.

Relationships with external states and TNCs

Trade blocs at times court controversy. While promoting free trade among internal partners, they are also vehicles for protectionism against imports from foreign nations and rival regions. The 2006 'bra wars' fiasco ranks as one of the more memorable examples of this practice in recent years. Eighty million items of Chinese-made underwear were blocked from entry at all EU ports by customs officers after official annual quotas were exceeded. (The quotas were designed to limit the numbers of garments entering the EU in order to protect local textile manufacturers.)

The global economic turndown of 2008 led rapidly to a doubling of similar trading restrictions. The most affected goods were agricultural products, steel, vehicles, chemicals, clothing, plastics and rubber (at one point, the USA placed a 35% duty on imports of Chinese tyres). To stay clear of such swingeing costs, manufacturing TNCs sometimes explore alternative business strategies. For instance, rather than exporting cars directly into an overseas trade bloc, another option for vehicle manufacturers is investment in assembly plants ('transplants') within the trade bloc or overseas market area. In this way, finished product import taxes and quotas can be bypassed entirely.

TNCs will seek out a **least-cost location** for production, factoring in transport and labour costs, and any local availability of host-state grants. Here are two examples:

- Since 1986, the Japanese firm Nissan has made cars for the EU market in Sunderland. It was originally attracted there by the availability of generous UK regional development aid and assistance.
- After EU enlargement in 2004, some American, Japanese and South Korean firms re-investigated production costs and subsequently decided to relocate their manufacturing from western to eastern Europe. South Korea's Samsung immediately moved its EU base of operations from Spain to Slovakia.

Managing the threat of sovereignty loss in Europe

Since the 1950s, the EU has grown from a six-nation Common Market to the point where it includes almost one in seven of the world's states (Table 7.2). Its 27 members now account for one-fifth of world economic output, giving the EU real global power. However, the speed of the grouping process has demanded careful and sensitive management. Past nation-building enterprises have brought bloodshed to the continent; during the twentieth century many millions of European lives were sacrificed to geopolitical struggle (the Second World War alone caused some 40 million deaths in Europe). Given such high prices paid for sovereignty in the past, a nation's right to self-determination may be something not to surrender lightly.

Among the electorate of each European country today a range of views is held about supranational integration and loss of state sovereignty. Groups such as concerned patriots, Second World War veterans, gap-year students, sun-seeking pensioners and homeowners in migration hotspots will most likely have varying and diverging perspectives on EU growth. Many European citizens and politicians fear and oppose a trajectory that is seen as tending over time towards the evolution of a single federal state, possibly modelled on the USA. The key flashpoints for debate are as follows:

- **Self-determination** This is measured as a state's continuing right to take important political or military action unilaterally. Even pro-European political parties have reservations about any suggestion of a future EU Constitution that could pave the way for a common foreign policy (thereby making it difficult or even illegal for any single state to act independently, as the UK did in supporting the US invasion of Iraq in 2003).
- **Financial insecurity** Corporate decision-making creates losers as well as winners within trade blocs (as seen when Electrolux moved its production wing from Sweden to Hungary). In recent years, TNCs have repositioned so many of their pieces on Europe's chessboard that local communities are left feeling unsure as to whether or not they are, on balance, beneficiaries.
- **Migration** An estimated 1.4 million eastern Europeans moved to the UK between 2004 and 2008, provoking protest not just from the British National Party (BNP) but also from moderate politicians.

Table 7.2 EU timeline

Year	Events
1950	Schuman Plan proposes a European Coal and Steel Community (ECSC).
1952	ECSC is created.
1957	Treaty of Rome establishing the European Economic Community (EEC) is signed by Belgium, France, West Germany, Italy, Luxembourg and the Netherlands.
1958	EEC comes into operation.
1962	Common Agricultural Policy (CAP) is agreed.
1967	EEC powers extended, becomes European Community (EC).
1968	Customs union completed.
1973	UK, Ireland and Denmark join the EC.
1979	European Monetary System established; first direct elections to the European Parliament.
1986	Spain and Portugal join the EC.
1993	European Union (EU) is established as the Maastricht Treaty comes into force.
1995	Austria, Finland and Sweden join the EU, which now has 15 members.
2002	Euro notes and coins come into circulation in 12 of the 15 EU member states.
2004	Ten new member states join the EU — Cyprus, Czech Republic, Estonia, Hungary, Latvia, Lithuania, Malta, Poland, Slovakia and Slovenia; EU Constitutional Treaty agreed.
2007	Romania and Bulgaria join the EU.
2008	Kosovo declares independence supported by EU.
2009	The Lisbon Treaty is ratified, allowing the election of the European Council's first president.

The last of these issues — immigration — can be a vexatious current affairs arena for politicians to enter into. Each state government attempts to negotiate a middle course between encouraging labour immigration (if deemed to be in the national economic interest) and being seen to curb it (should popular voter concerns arise). After negotiations for the accession of ten new member states were completed in December 2002, some existing EU nations introduced measures to limit the international movement of new eastern European migrants until 2011. Portugal limited work permits to 6500, while Germany banned migrant workers in many sectors of its labour market for the full 7-year period allowed under accession rules. The UK took a more relaxed attitude, which resulted in over a million arrivals between 2004 and 2007. However, the UK parliament has strengthened measures to limit migration flows from outside the EU with the roll-out in 2008 of a new five-tier immigration points system.

The reaction of political parties and civil society in Europe

An important political aid for national governments navigating the often contested route towards supranational governance is the **referendum**. Countries joining the EU usually do so following a public vote in favour of membership; the population of Norway rejected accession in 1995 following a referendum. European states have sometimes held further referenda in response to proposed major changes in the functioning or management of the EU, such as the ratification of the Lisbon Treaty and the election of Europe's first president, Herman van Rompuy, in 2009.

A referendum result in favour of joining the EU provides state leaders with a clear mandate to surrender, or partially surrender, certain previously-held sovereign rights in exchange for new membership privileges. However, unless the vote has been overwhelmingly positive, a sizeable minority of the electorate in each member country will remain hostile to EU governance — usually backed by sections of the media. Some may hold out hope for a future reversal of the decision to join, and in most countries 'Eurosceptic' political parties, such as UKIP (the UK Independence Party), continue to command significant public support. British Eurosceptics greeted with dismay the recent news that the national birthrate for the UK had risen to 1.96 children per woman, its highest level in over 30 years, when it emerged that much of the increase had come from eastern European women now living in the UK, whose total birthrate is relatively high at 2.5 children.

Figure 7.5 Polish beer on sale in a London grocer's

Simon Oakes

In contrast, many other British people have welcomed European migrants and closer ties with continental Europe. For instance, small shopkeepers throughout the UK have visibly courted Polish custom (Figure 7.5). Mixed marriages and increased numbers of Polish children enrolled in primary schools speak volumes about cultural inclusion. So too do reports of English police officers taking Polish classes in their leisure time; or accounts of Anglo-Polish rock bands playing at London music venues.

In conclusion, growth of the EU — in common with the expansion of regional trade blocs elsewhere in the world — has impacted on the lives of state citizens in significant ways. Perhaps the most profound change of all has been a constantly increasing correlation between the growth in each nation's level of global economic activity and the rising level of concern expressed by many of its citizens that their unique national identity is threatened. This theme is further explored in Chapter 8.

8 Local responses and challenges to globalisation

Globalisation and glocalisation

The practice of **glocalisation** developed from the local sourcing of parts by TNCs when establishing branch plants overseas. It involves using components from regional suppliers to assemble 'global products' closer to the local markets where they will be sold. As levels of consumption for goods and services have risen in emerging economies such as India and China (in the past viewed solely as sites of production for Western markets), firms have increasingly customised their products for sale in these newly appeared consumer markets as a global sales strategy. Glocalisation helps a firm to gain acceptance as part of the local business community, while customising the product to meet indigenous tastes encourages its local diffusion. Working in partnership with Uni-President China Holdings Ltd, the European TNC GlaxoSmithKline rebranded its energy drink Lucozade for the Chinese market. The product's new local name translates as 'excellent suitable glucose'. The drink has also been reformulated with a more intense flavour.

Glocalisation has become an economic, political and cultural strategy that informs any ambitious firm's decision to enter the global marketplace. The geographer Peter Jackson believes that for all the corporate energy that has gone into creating 'global' marketing messages, cultural homogenisation as a result of worldwide diffusion of TNC goods and services is still far from being a foregone conclusion. Jackson argues that producers, rather than rolling out their existing products across a geographically undifferentiated market, will adapt their global brands to meet any variety of local conditions. This results in an apparent paradox, whereby big-brand globalisation in fact requires companies to sensitively adopt a variety of 'localising strategies'.

Take the case of global media providers such as Disney, MTV or the BBC. Televisions, films and books have long been translated or subtitled for foreign markets: in the 1930s, classic American comedy duo show

Laurel and Hardy was repackaged as *Dick und Doof* in Germany and as *Cric e Croc* in Italy. Today, organisations such as the BBC take the process to a new level, both in terms of how properties are licensed and produced; and also in the multiple ways that stories, games or other entertainment formats such as 'reality' television shows are comprehensively redesigned, rewritten or re-filmed to feature local characters in local settings, while not compromising the totality of the global brand identity. Production company Endemol UK has licensed the rights to its *Big Brother* franchise in nearly 70 countries, sometimes even taking the unusual step of renaming the show (in Canada it is called *Loft Story*).

Figure 8.1 Instances of corporate glocalisation

McDonald's Corporation	MTV (Music Television)	The Walt Disney Company
For its sales in India, 98% of McDonald's ingredients and paper products are sourced from within India itself. The Indian product development team has also generated innovative items like the popular McAloo Tikki burger, the Chicken Kabab burger and the Maharajah Mac (a non-vegetarian mutton burger), all served with small fries. McDonald's is run as a franchised partnership in India's major cities. For instance, Connaught Plaza Restaurants runs nine branded McDonald's restaurants in New Delhi. This franchising model is followed in all of the 118 countries where McDonald's operates outside of its home market in the USA.	MTV Networks uses a **360° strategy** involving 'full spectrum' marketing to maximise its global audience. It has grown over time through acquisitions and mergers with existing companies, as well as by setting up new regional service providers such as MTV Base (available since 2005 to around 50 million viewers across 48 countries in sub-Saharan Africa via satellite). In 2008, new channel MTV Arabia began broadcasting to Egypt, Saudi Arabia and Dubai. Two-thirds of the Arab world is younger than 30; many of these young people are fans of cutting-edge music, especially hip-hop, which MTV Arabia now specialises in broadcasting.	In 2009, Disney released its first Russian film, *Book of Masters*. Based on a Russian fairy tale and produced using local talent, the film reflects Russia's new position as one of the world's fastest-growing cinema markets. Disney's Russian publishing unit sold 7 million books during 2008 and produces 13 magazines with a combined circulation of over 1 million. *Roadside Romeo* (2008) is the first film that Disney has made inside India. It is also aimed at local audiences and uses home-grown animation put together by Tata Elxsi's Visual Computing Labs (VCL) unit. A co-production with India's Yash Raj studios, this film tells the story of a dog living in Mumbai.
'We have to keep our ears to the ground to know what the local customer desires...(it's) key to our worldwide functioning.' (Amit Jatia, MD of McDonald's India West and South)	'We will respect our audience's culture and upbringing without diluting the essence of MTV.' (Bhavneet Singh, MD of MTV Emerging Markets)	'There is great interest and pride in local culture. Even though technology is breaking down borders, we're not seeing homogeneity of cultures.' (Bob Iger, Disney CEO)

The strengthening of Latin American, Asian and Middle Eastern economies has prompted an explosion of media programming aimed at cash-rich young people in places previously sidelined by TNCs. Demography is an important factor here: most emerging economies have youthful populations due to the persistence of high birthrates. In India alone, there are 100 million children aged ten or under. This fact has not escaped the attention of TNCs. However, children and young adults are often selective consumers: products and services must be carefully researched and refined if they are to win the hearts of local youth (Figure 8.1). TNCs may need visibly to show respect for different local customs in order to guarantee a high level of success beyond the margins of their home markets. In the Middle East, MTV avoids unedited screening of more controversial music videos that play uncensored on sister Western channels.

Collaboration with local talent spells financial success for TNCs in other ways too. Rather than simply exporting American or European goods and culture worldwide, global firms can mine new local territories for fresh music, food or fashion ideas that are fed back in turn to core markets. Thus the Japanese 'manga' style of cartoon art tunnelled its way back into America's mainstream via Disney and Marvel studios during the 1990s. In the 1980s, African music was enthusiastically embraced by Western musicians including Paul Simon and Peter Gabriel, a trend that continues today, with European artists such as Franz Ferdinand and Mika grafting polyrhythmic African beats to their songs.

Music from all over the world has gained exposure both through formal means (the major media corporations) and informally via electronic distribution channels such as YouTube. The internet brought success to Mexican acoustic guitar duo Roberto y Gabriela, an act inspired by classic Western rock acts like Pink Floyd but whose music also clearly displays Latin American rhythmic and melodic traits. Now signed to a global record label, their music has been distributed, sold, digitally downloaded and listened to worldwide. Completing the cultural circuit, Western guitarists who listen to Roberto y Gabriela may now choose to infuse their own locally learned style with a little Mexican colour.

Evaluating glocalisation

Do widespread practices of glocalisation and hybridisation sufficiently redress accusations of cultural imperialism (Chapter 5), thereby rendering TNCs relatively harmless in the grand scheme of things? The answer to this question hinges upon perspectives and values. It can be argued that a McDonald's burger sold in India is a vehicle for the Americanisation of Asia, irrespective of the fact that extra spice is added and a beef substitute used (in deference to local tastes and beliefs). Consider also the 'gatekeeping' role that MTV's international executives may play in selecting local talent to slot into the predominantly Anglo-American output of their regional satellite stations. The most favoured local artists will be those who complement mainstream commercial programming. More diverse indigenous musical expressions are more likely to be left languishing on the sidelines.

The hybridised nature of goods and services sold at local markets by TNCs must also not blind us to the fact that a more fundamental sweeping change is still occurring, namely the commercialisation of food, fashion, drink and other elements of people's lifestyles. Globally dominant firms are responsible for spreading the consumerist ethos at a planetary level. The twentieth-century philosopher Antonio Gramsci used the term 'hegemony' to describe the exercise of power on this scale. Powerful TNCs are shaping a new, popular consensus that sees their products as valued and desired by people in newly emerging economies such as Brazil or India. Long-established uncommercial forms of leisure, play and entertainment are rapidly giving way to consumer aspirations among growing sections of the populations living in these places.

Figure 8.2 How global TNCs 'enrol' local people as consumers in emerging economies

The products of TNCs are **promoted** in the local marketplace using advertising (on television channels, in newspapers and on billboard hoardings where people can clearly see them)

Message sent

'Messages' are **encoded** into adverts carefully designed for each local market. TNCs want their products to be viewed as:

- an important part of everyday local life
- things that everyone aspires to consume as part of a more commercialised lifestyle

Message received

'Messages' are **decoded** by local people in the way that the TNC intends, especially when the glocalisation strategy is usefully employed:

- people desire to experience the product
- by adopting the product, people are made to feel they are making economic progress

Message created

Transnational corporations (including Disney, Coca-Cola, McDonald's, MTV, Wal-Mart) may:

- seek to **recruit** consumers in new markets
- be owners of digital infrastructure that helps them advertise goods and services

Building **cultural hegemony** ('the manufacturing of consent')

Message acted on

People in emerging markets are now successfully **enrolled** as TNC product consumers. They willingly consent to buy products that are non-essential, or even unhealthy (such as burgers or cigarettes)

Widespread adoption of global brands such as Starbucks, McDonald's and Disney suggests that these products are greeted with enthusiasm in new markets, where their uptake reflects free choice rather than coercion (after all, no one has taken a gun to force anyone in India or Brazil to eat a Big Mac). Yet Gramsci has argued that the decisions made by consumers are strongly influenced by powerful subliminal messages carried by adverts for branded products. These penetrate deeply into local markets everywhere thanks to glocalising strategies (Figure 8.2).

Figure 8.3 Early 'glocalised' advertising: Sierra Leone, 1968

The worldwide spread of Western capitalism has also been critically analysed by Noam Chomsky, who describes the consumerist model's unarrested global success as 'the manufacturing of consent'. According to this viewpoint, any impression that local people are equal partners in a 'cultural conversation' taking place with TNCs is false. Global food and drink conglomerates like McDonald's and Diageo merely pay lip service to local cultures by 'tweaking' the taste of their generic products, or the slant of their advertisements, in order to drive sales higher in differing markets (Figure 8.3).

Global diaspora groups and local community interaction

TNCs are not the only global players to adopt 'localising strategies' *in situ*. Mirroring economic glocalisation are processes of cultural change and exchange experienced by members of global diasporas who find themselves interacting with different populations in a range of local settings. For instance, many young British Asians have opted to embrace certain mainstream British culture traits while also choosing to retain and honour aspects of their

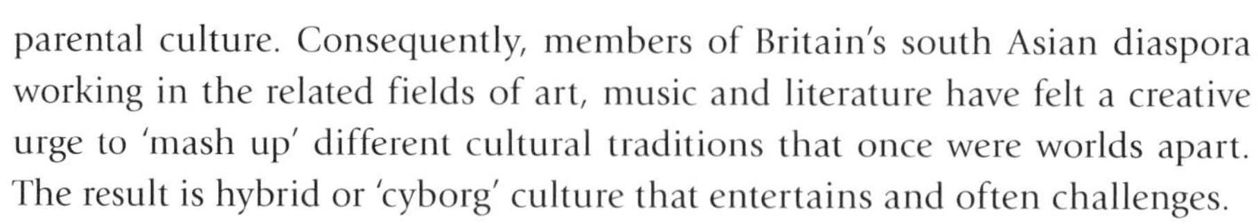

parental culture. Consequently, members of Britain's south Asian diaspora working in the related fields of art, music and literature have felt a creative urge to 'mash up' different cultural traditions that once were worlds apart. The result is hybrid or 'cyborg' culture that entertains and often challenges.

- The films *East is East* and Gurinder Chadha's *Bend it Like Beckham* have attracted large multicultural audiences. Chadha's overtly 'melting-pot' film title *Bride and Prejudice,* and theatre director Deepak Verma's Bollywood musical restaging of *Wuthering Heights,* are deliberate interventions with iconic cultural forms.
- London's Jay Sean (whose real name is Kamaljeet Singh Jhooti) has brought elements of south Asian music into Britain's mainstream. Music producer Diamond 'Swami' Duggal used Asian bhanghra-tinged orchestration to help make Shania Twain's album *Up!* a bestseller.

Figure 8.4 British students perform a 'hybrid' dance that blends 'Eastern' and 'Western' influences

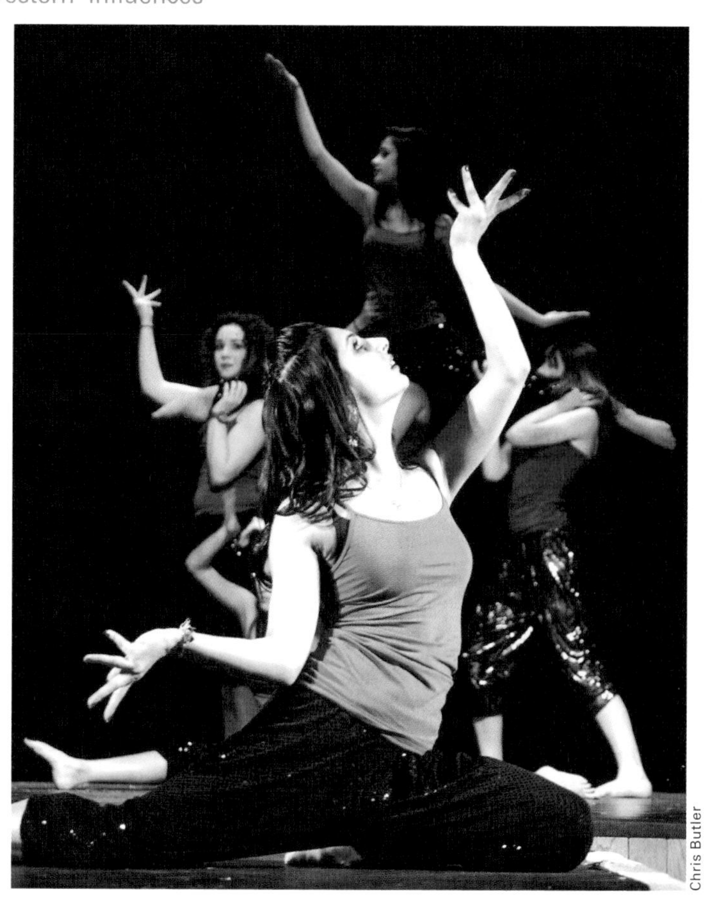

Chris Butler

- The postwar influence of south Asian food on UK cuisine cannot be overstated. Essex-born television chef Jamie Oliver owes much of his success to what he calls the 'Aladdin's cave' of restaurants in London's Brick Lane.
- Author Salman Rushdie has explored the changing relationship between global cultures of 'East' and 'West', though the Mumbai-born Booker prize-winner's 1988 novel *The Satanic Verses* caused offence to many Muslims through its blurring of fiction with facts about the Islamic faith.

At a local level, UK schools and colleges are important spaces where personal identity is negotiated and explored by young British Asians. At Bancroft's School in Essex, a large mixed student body of Muslims, Sikhs, Hindus and Buddhists collaborate each year in the Taal festival (Figure 8.4). Traditional and modern music and dance combine to provide a complex cultural expression of what it is like to grow up with one foot in London life and another in Asian family culture.

In contrast, other members of global diaspora groups — both in London and other world cities — choose to uphold distinct cultural identities. Some Muslim women in the UK adhere to strict dress codes and wear the traditional face-covering niqab veil in public (described by Labour politician Jack Straw in 2006 as a 'visible statement of separation and of difference'). In France, continued use of the body-covering burqa by a small minority of women of north African descent has attracted criticism from some politicians, including President Nicolas Sarkozy. Not every migrant, or descendant of migrants, wants his or her own culture to hybridise with that of their place of residence (which may, in any case, be viewed as no more than a temporary workspace). Diaspora members, like any group of people, vary considerably in their inclination to embrace changes.

Activity 1

Examine the evidence for cultural exchanges in your own school or neighbourhood community. Alternatively, analyse the characterisation of a popular soap opera, such as the UK television programme *Eastenders*, asking (a) what diaspora groups are represented and (b) how the show's scriptwriters portray the lives of 'global' groups living in a tight-knit local community.

Local resistance to globalisation

As previous chapters have indicated, globalisation may be criticised on multiple grounds as a process that is associated with economic injustice

(Chapter 4), cultural imperialism (Chapter 5) and unsustainable environmental impacts (Chapter 6).

Power to take action to resist any or all of these aspects of globalisation is invested in many different players at a range of scales and positions of responsibility. The heads of state of several countries, including Venezuela, Cuba and North Korea, are vocal opponents of the 'new world order'. So too are influential non-governmental organisations (NGOs) such as Amnesty International, Greenpeace and Christian Aid (Table 8.1). NGOs are part of **civil society**, along with private citizens, many of whom strongly believe that the poorly regulated actions of globalisation's most powerful economic players — such as banks and food companies — should be opposed, or at least have their excesses curbed and damaging effects ameliorated.

Political demonstrations and protests became commonplace during the late 1990s and 'noughties', often orchestrated by private citizens interfacing with new technology through social networking and microblogging sites such as Facebook or Twitter. Examples of key oppositional moments include:

- **Seattle World Trade Organization conference, 1999** Forty thousand protestors flooded the streets and voiced disapproval of WTO guidelines for global trade that, critics argue, deny poorer nations 'a level playing field'. Rioting resulted in a state of emergency being declared in Seattle.

Table 8.1 How players resist globalisation at a range of scales and for different reasons

Individuals and communities	Non-governmental organisations (NGOs) and pressure groups	Sovereign states
Consumers can refuse to give custom to high-profile high-street global brands like Starbucks, or may boycott polluting air travel	**Amnesty International** has opposed the negative impacts TNCs have had on the human rights of some local communities	**Venezuela** follows a tradition of Latin American socialism opposed to free-market globalisation under the leadership of Hugo Chávez
Critics of globalisation include powerful writers (Naomi Klein, Susan George) and publications (*Le Monde diplomatique*)	**World Social Forum** is an annual meeting in Brazil where committed individuals and NGOs campaign for global social justice	**North Korea** (under communist rule of Kim Jong-Il and his family) is seen as a 'rogue state' by the USA (the leading global superpower)
Townspeople living in the small English town of Todmorden aim to make it entirely self-sufficient in food production	**MigrationWatch UK** is an NGO that views current levels of international migration into the UK as growing 'out of control'	**Ethiopia** adheres to a calendar running 7 years behind the rest of the world (it celebrated its year 2000 on 12 September 2007)

- **Cancún (Mexico) world trade talks, 2003** South Korean farmer Lee Kyung-hae died after stabbing himself in protest against a continuing lack of reform for WTO global trade laws.
- **Copenhagen climate change conference, 2009** Many hundreds of campaigners were detained by police after 100 000 activists, some dressed as polar bears and penguins, took to the streets of Copenhagen while the world's political leaders failed to commit to a new binding agreement to tackle global climate change.

Activity 2

Study Table 8.1 and choose three examples to research further (one from each column). For example, you can find out more about Todmorden's 'Incredible Edible' campaign and about World Social Forum at these websites: **www.incredible-edible-todmorden.co.uk/** and **www.wsftv.net**

Taking action through 'local sourcing'

The phrase 'local sourcing' describes the practice of buying goods and services solely from local suppliers, thereby boycotting the use of extended global supply networks such as those favoured by most supermarkets. The clearest environmental benefit is believed to be a markedly reduced number of **food miles** for agricultural products. Providing stimulus for local production has other merits too. Greater regionalised agricultural activity can improve national **food security** in an era when climate change and growing global demand put mounting pressure on the world's food supplies. In developed nations, greater local sourcing of food can also boost employment in rural areas.

A cursory look at any supermarket aisle will yield no shortage of instances where imported food, drink and consumer goods have travelled excessive distances. Freshly picked vegetables like asparagus are routinely transported into Europe by air from South America. UK supermarket Sainsbury's sources charcoal briquettes for barbecues from South Africa. Setting light in a British garden to wood that has been burned once already — having also virtually travelled half the globe in a carbon-emitting container ship — surely ranks high on any scale measuring environmental folly.

Local sourcing wins plaudits from governments and is promoted by those businesses that have declared their commitment to carbon footprint reduction. However, some foreign agricultural imports into the UK — Spanish and north African tomatoes and flowers, for example — may in fact do less environmental harm than some local agricultural systems that are reliant on energy-hungry heated greenhouses (Figure 8.5). Blaming globalisation for increased carbon emissions can sometimes be a misreading of a more fundamental issue. After all, globalisation is not directly responsible for British people wanting to buy large amounts of flowers in midwinter: the root cause is commercialisation of the Valentine's Day festival. This drives a demand that can be met only by flying to market in the UK flowers from Equatorial and Southern Hemisphere regions, or through greater local use of heated greenhouses.

Opting to buy fewer non-essential goods such as flowers in the first place — irrespective of their local or global origin — would be a more effective way for wealthy societies to mitigate climate change than local sourcing. The

Figure 8.5 Resisting globalisation through the 'local sourcing' of food in the UK: how clear are the environmental benefits?

How can local sourcing help reduce CO_2 emissions?

The global food industry has created a large carbon footprint over time. Increased consumer demand for food choices at UK supermarkets has led to complex changes in the pattern of sourcing of produce to provide a year-round supply of fresh and sometimes exotic fruit and vegetables — many of which will travel thousands of miles by aeroplane.

Airfreighted goods are especially polluting. Transport of food by air has the highest CO_2 emissions per tonne of any mode of transport. Although airfreight into the UK accounts for only 1% of travel distances (in tonnage) for food sold in supermarkets, it is responsible for 11% of national food transport CO_2 emissions.

Why do the claimed benefits of local sourcing require careful evaluation?

Local food may travel long distances too. Food grown for supermarkets inside the UK may still be routed through supermarket regional distribution centres, and can travel long distances on motorways in heavy goods vehicles (HGVs). In addition, consumers sometimes travel long distances by car to out-of-town shopping centres in order to buy this 'local' food.

Local transport can be inefficient. Large aeroplanes and HGVs may travel long distances but they are also efficiently loaded vehicles, which reduces the impact per tonne of food. Locally sourced food may have travelled shorter distances but often in much smaller vehicles, meaning that CO_2 emissions per tonne are relatively high.

Local food production systems can be energy-intensive. It can be more sustainable (at least in energy efficiency terms) to import tomatoes from Spain than to produce them in heated greenhouses in the UK outside the summer months.

There may be important social benefits to buying global food. Many developing nations are dependent on food exports to countries such as the UK. It would hurt Kenyan farmers if all UK consumers ceased buying Kenyan runner beans because of the high food miles attached to them.

downside of this, however, is that it would be enormously damaging not only for global trade but also for the livelihoods of thousands of people in poor countries like Kenya.

Switched-off places and peoples

In Chapter 1, degrees of global integration were analysed for different nations. The **least developed countries** (LDCs) are a grouping of the world's poorest states that are, in many cases, involuntarily excluded from highly profitable global interactions taking place elsewhere. Around 50 such nations are described by some academics as 'Fourth World' countries to emphasise the globally 'switched-off' conditions that prevail in places like Somalia and Eritrea, where the economic outlook remains bleak. (A few, including Sudan, Afghanistan and Chad, have even been described by political commentators as 'failed states', i.e. without any effective government.) The majority of citizens in these countries lack any positive connection with global flows of information and money.

Suggested reasons for the systematic failure (economic, political or both) of some states and groups of people to benefit from globalisation have featured at intervals throughout this book. In summary, they include:

- **failure to develop human resources** (in postcolonial Sierra Leone, a whole generation was left uneducated when children as young as seven were conscripted as soldiers in the 'blood diamonds' conflict)
- **protectionism and agricultural subsidies** by rich trade blocs (the £30 billon-a-year EU agricultural subsidy regime drives down prices paid by supermarkets to African farmers who cannot compete with the low prices for EU produce made possible by such payments)
- **mismanagement of financial flows**, including IMF loans, international aid and trade (Mobutu Sese Seko, former president of Zaïre — now the DRC — accumulated massive personal wealth by diverting borrowed funds, foreign aid and revenues from state-owned mineral companies)
- **isolationist policies** (Zimbabwean leader Robert Mugabe's aggressive criticism of powerful Western nations and their organisations has deterred foreign investors who also judge the territory to be lacking enforceable property rights and a reliable legal system)
- **military intervention** (Afghanistan was invaded by the Soviet Union in 1979 and again in 2001 by the USA and allies as part of the 'war on terror')

Any of the above can contribute to **capital flight** — the financial flows that connect a place to the global economy fail when investors lose confidence,

assets are devalued, exchange rates fall or a nation's currency is devalued as a result of domestic strife.

Activity 3

Evaluate the relative importance of the factors responsible for some places remaining 'switched off' from globalisation. Using a reputable website, such as that of BBC News, research case studies suggested here (Sierra Leone, the DRC, Zimbabwe or Afghanistan). Examine your evidence for signs of a primary reason why some places have failed to benefit from globalisation.

The last remaining non-globalised people

Few places today remain entirely disconnected from globalising processes. Few people do not participate, if only very occasionally, in long-distance trade or information exchanges. The only truly non-globalised societies are, arguably, **uncontacted tribes**. These are social groups who have yet to make contact with the outside world (Figure 8.6) and whose existence may only be known through satellite imagery. South America's Amazon rainforest (still an enormous wilderness region, despite ongoing road-building programmes) is home to the world's largest number of uncontacted tribes (Figure 8.7).

Figure 8.6 An uncontacted Amazon tribe: Survival International estimates that there are over 100 uncontacted tribes worldwide

© Gleison Miranda/FUNAI (Image supplied courtesy of Survival International www.survivalinternational.org/)

Figure 8.7 The location of some of the world's remaining uncontacted tribes — global society's most 'switched-off' peoples

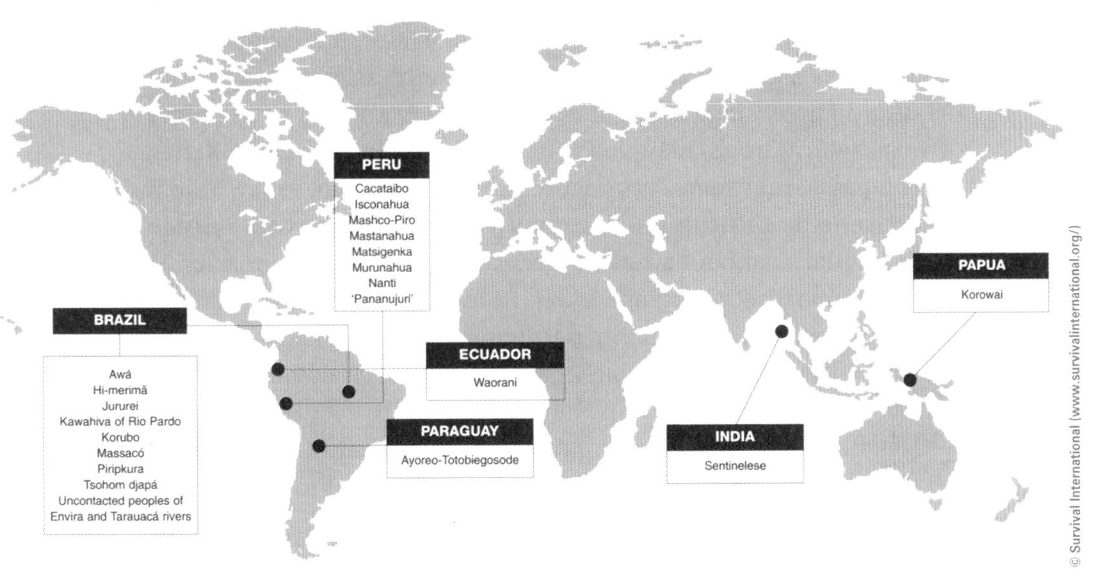

© Survival International (www.survivalinternational.org/)

These uncontacted tribes belong to a wider group of **indigenous people**. Indigenous people are ethnic groups who have enjoyed continuous occupation of certain places for long periods of time and pre-date the arrival of more recent migrants. Some small pockets of indigenous people have remained cut off from the outside world up to the present day — quite an achievement in the era of modern globalisation. The explanation for their isolation lies with the small numbers involved and their habitation of peripheral wilderness areas that have never been mapped at ground level or photographed aerially at high resolution.

In northeast Brazil, ground level is in any case obscured by the lush, dense tropical rainforest canopy, helping some small settlements to go undetected by aircraft. The Brazilian government has estimated that between 30 and 53 separate ethnic tribal groups live in this region. Since the 1980s, state agency FUNAI (Fundação Nacional do Índio) has adopted an official position whereby it attempts to map the territory of uncontacted tribes but does not actually make contact unless absolutely necessary. This is due to the many risks involved, including possible spread of diseases (such as measles) to indigenous people who lack immunity. This policy is unlikely to have long-lasting success due to increased external forces intruding on tribal habitats. These forces include illegal or unregulated and criminal activities brought by economic globalisation, such as:

- oil companies constantly on the lookout for fresh reserves beneath Amazonia (Perenco, the Anglo-French oil TNC, has sent hundreds of workers deep into tribal territories)
- loggers penetrating further into Amazonia in search of valuable timber, or to clear forest for globally important crops such as soya
- clashes between international cocaine traders and government anti-drug troops, which drove Colombia's Nukak tribe from its forest isolation

It is possible that some supposedly 'uncontacted' tribes may have isolated themselves after difficult or dangerous early experiences with outside forces such as settlers, rubber barons, loggers or evangelical missionaries. Viewing their non-globalised existence today from a modern urban perspective, it is perhaps easy to over-romanticise their existence as 'innocent', 'simple' or 'blameless' (given these people's negligible ecological footprint compared with more globalised societies). However, theirs are lives frequently cut short by treatable injury and disease, problems in childbirth or murder in inter-tribal warfare. As with other debates relating to globalisation, there are two sides to this 'switched-off' story that we need to consider critically.

Activity 4

Research more facts about the last truly 'switched-off' societies on Earth at **www.survivalinternational.org/** and **www.survivalinternational.org/uncontactedtribes/**. Do you agree that there is a moral imperative to leave these tribes alone — or instead to inform them of the opportunities modern healthcare and education bring?

9 Global futures

'De-globalisation': the global credit crunch of 2008–09

The 1990s and early 'noughties' were a golden age for globalisation. New technologies transformed world trade, while many previously communist economies chose to realign themselves with global capitalism. In contrast, the collapse that began in 2008 brought a sudden end to the era of confidence and expansion, and triggered an abrupt decoupling of global interconnections between businesses, places and people. Financial markets lapsed rapidly into panic when the value of assets held by major world banks plummeted. The interdependent nature of the global economy ensured that the crisis spread instantly and extensively. Instead of creating wealth, **financial connectivity** contributed to its annihilation on an unprecedented scale. In the 12 months after September 2008, world GDP fell for the first time since 1945. Globalisation was discovered to have a 'reverse gear' (Figure 9.1).

This was not the first postwar occurrence of reduced economic growth nor of a slowing-down of global flows such as migration and investment in response to world events. Precursors include the major OPEC oil crisis of the 1970s and, more recently, a series of less serious challenges (Table 9.1). However, the scale and impacts of the global credit crunch were beyond the worst expectations of many critics. World trade fell at roughly twice the rate experienced during the Great Depression of the 1930s. Several major financial institutions failed and a near-total collapse of global economic confidence occurred, with share prices

Figure 9.1 World GDP growth for key players, 2000–14 (IMF estimates)

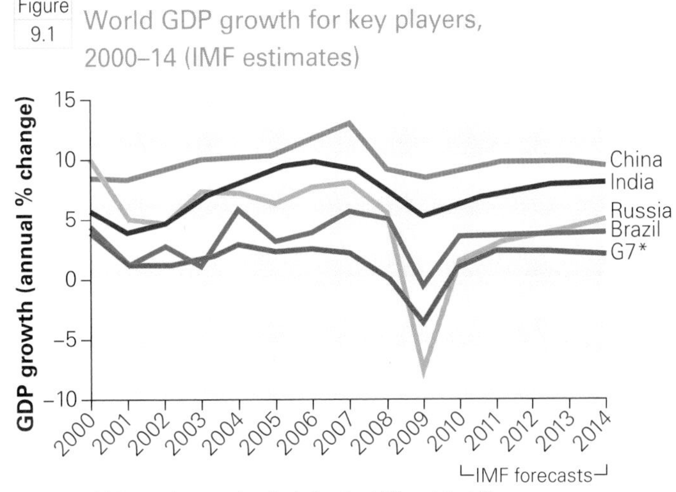

*Advanced economies, including the USA and the UK

Table 9.1 The 'noughties': a decade of rising global concerns

Year	Crisis	Symptoms
2001	'Dot-com' bubble bursts	Over-inflated valuation of many hi-tech companies, notably internet operations, results in a colossal 'bust' as global share prices collapse
	9/11 attack on USA	Terrorists fly two hijacked airliners into the World Trade Center 'twin towers' in New York, ushering in a new era of global insecurity
2003	Sars outbreak	A global spread of this pneumonia-like condition is initially feared after its outbreak in China; Canada experiences a major scare
2008	Food crisis	A dramatic rise in global food prices results in shortages and riots breaking out across the world, from Egypt to Haiti
	New oil crisis	Between 2003 and mid-2008, oil prices rise by almost 500%, leading to heightened geopolitical concerns over oil security
	Global credit crunch	After early warning signs in 2007, a dramatic reassessment of the assets held by many leading international banks results in the worst global economic slowdown since 1945
2009	Swine 'flu	Another global health scare erupts, this time over fears that the H1N1 influenza virus will spread globally, following an outbreak in Mexico City

tumbling across all stock exchanges. Many national economies entered a period of recession from which some had still not emerged by the start of the 'teens' in 2010 (Figure 9.2).

Figure 9.2 How some nations fared economically in 2009 (the year after the credit crunch)

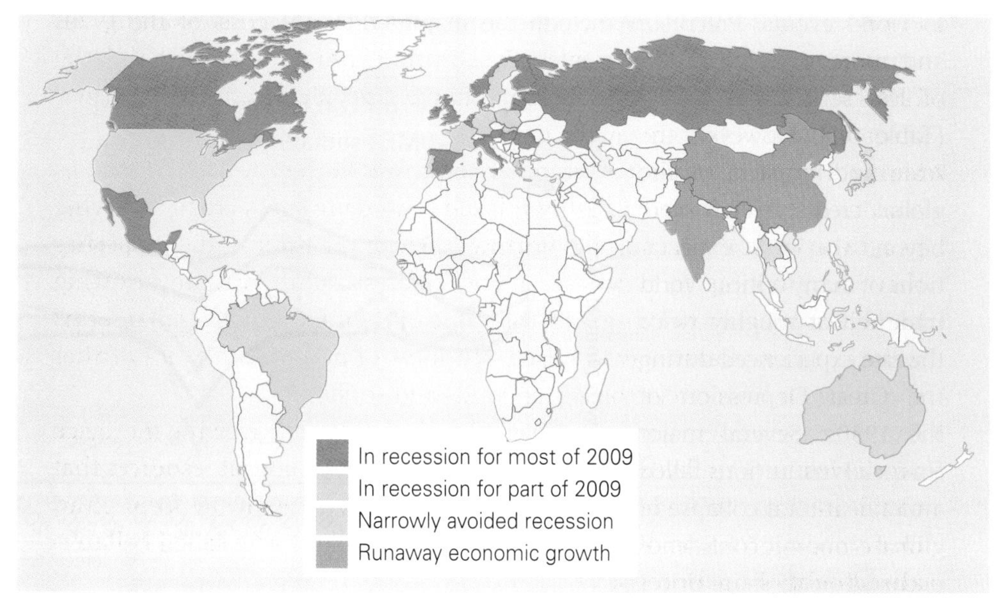

Activity 1

Study Figure 9.1 and compare the trends shown for emerging and advanced economies after 2000. Suggest reasons for similarities and differences you identify.

The origins of 'globalisation in reverse gear'

The Asian Development Bank estimated that financial assets worldwide shed more than US$50 trillion in value during 2008 as a result of the global credit crunch — a figure of the same order as total annual global economic output. In the UK and USA, the market capitalisation of banks plummeted by 80%. The weakly regulated pursuit of wealth, as practised at a range of scales by individuals, institutions and national governments (the last keen to boost GDP through successful performance of their own national financial markets), created a major economic 'boom'. In line with previous periods of speculation and asset bubble growth, this boom finally gave way to a colossal 'bust'.

This was a complex, multi-strand and multi-scalar crisis, the explanation of which requires analysis of several interlinked factors. These include: global imbalances in levels of wealth creation; changes in the regulation and operation of money markets; the growth of a very high-risk-based lending culture in the USA and other developed nations; and the runaway aspirations of ordinary consumers living in these countries. The following points outline events surrounding the landmark collapse in September 2008 of Lehman Brothers (a major US-based investment bank):

1 Prior to September 2008, American real-estate agents working on commission had pushed up sales of 'sub-prime' property mortgages in the USA (by offering large housing loans to poor people, irrespective of their income and ability to repay the debt).
2 Similar trends were followed in other countries. In the UK, lenders such as Northern Rock adopted a highly relaxed attitude towards credit, allowing people on ordinary incomes to borrow 125% of the value of the properties they were looking to purchase, at low interest rates. For many existing homeowners, borrowing against the value of their homes on similar terms was a cheap and easy way to raise large sums of money to finance further spending on high-street consumer goods and services.
3 Banks designed new ways to convert the enormous and growing mortgage debt owed to them by homeowners into **bonds** — financial resources that could, in turn, be traded with other firms, sometimes in the form of an insurance deal. Some of the most well-known bonds were called **collateralised debt obligation** (CDO).

4 Trade in CDO bonds was so successful that it fostered a widespread sense of financial security in the banking industry. Any risk was perceived as being well spread on account of the many different organisations and individuals investing in bonds. This encouraged even greater levels of lending and even fewer calls for prospective borrowers to prove repayment capability. By 2007, Northern Rock was lending three times more money, per pound of its own assets, than in 2004.

5 The turning point came when defaults (failure to make repayments) on US sub-prime mortgages started to rise, and US house prices fell. Bond values likewise shrank. In a culture of fear, confidence fell further and sales of financial products plummeted. Many banks faced sudden bankruptcy after finding their assets were worth far less than previously calculated.

6 In September 2008, Lehman Brothers filed for bankruptcy. Following its multi-billion-dollar collapse, facts emerged showing that Lehman's had lent 35 times more money than it held in assets (meaning that it owned only $1 out of every $35 registered in its accounts). This excessive **leverage** had proved unsustainable.

7 A sudden evaporation of financial liquidity (i.e. the ready availability of bank loans) saw global trade go into freefall. House prices collapsed. Corporations and homeowners postponed major spending decisions (such as purchasing cars or foreign holidays), leaving manufacturers with warehouses full of unsold goods and without cash to pay staff wages. This negative multiplier effect saw G7 economies shrink at an annual rate of 8% in the first quarter of 2009.

Identifying a root cause for the credit crunch

Many commentators believe that blame for this epic mismanagement of global capital lies with people working in the financial sector, in particular with New York and London bankers who were all too ready to exploit the openness of the market and become involved in reckless speculation and lending. The weak regulation in place was a result of key government decisions taken during the 1990s, when politicians deregulated banks, thereby, *inter alia*, allowing them to trade high-risk loans on to third parties via bonds.

Not everyone blames Western governments and corporation chiefs alone for the global credit crunch, however. According to an alternative viewpoint, the fundamentally unbalanced yet interconnected nature of recent global growth was a more important root factor. In particular, China developed an enormous trade surplus after 2000. This left the Chinese state with a giant savings glut held in US dollars, which subsequently depressed borrowing costs (interest rates) for US and European banking customers and fuelled the cheap loans and spending bonanza. It can be argued that this financial connectivity caused the severity of the collapse (Figure 9.3).

Figure 9.3 Financial connectivity as the root cause of the global credit crunch

The geography of 'decoupled connections'

The events of 2008–09 can be interpreted as a crisis of globalisation (Table 9.2). It was fostered and transmitted by the rapid and deep financial integration of national economies that vary greatly in their nature and functioning— the

Table 9.2 Impacts of the global credit crunch at varying geographical scales

Impact	Scale
As emerging economies prove to be more resilient than developed economies in the global economic downturn, a perceptible shift in power and influence occurs, symbolised by the expanded G20 group rivalling and perhaps overtaking the G7/8 group as the nexus of global decision-making.	**Supranational**
Some economies are plunged into crisis, notably Iceland, Ireland, Spain, Greece and Dubai (UAE). Major economies such as the USA and UK are left with national debts that are unlikely to be paid off for at least a decade or more.	**National**
A global slump in consumption causes factory closures or workplace rationalisation. Regional economies that are heavily dependent on manufacturing, such as the US state of Michigan and China's Guangdong province, initially experience a sharp rise in job losses.	**Regional**
Some UK city councils such as Westminster lose funds that were invested in bankrupt Icelandic banks. UK high streets lose major names such as Woolworths, while McDonald's exits Iceland's capital city of Reykjavik.	**Local**

most significant macroeconomic contrast being that between the 'surplus' country of China and the 'deficit' nation of the USA. More widely, patterns of outsourcing built up over recent decades (during the golden age of globalisation) left the global economy highly vulnerable to any fall-off in demand. For instance, hi-tech products such as laptops contain hundreds of parts sourced from upstream suppliers for final assembly, often in export processing zones sited strategically around the world. If a major assembly firm collapses, then businesses that supply component parts find themselves instantly at risk too.

The worst-affected places therefore experienced a decoupling and dis-aggregation of their industrial linkages. Interconnected firms became vulnerable to the 'domino effect' that brings damage to entire assembly 'clusters' during a downturn. In the UK, aerospace firms in northwest England and chemical firms on Teesside operate in clusters, with firms supplying one another with vital parts or services. So when Dow Chemical Company pulled out of Teesside during the global credit crunch, this act immediately triggered the closure of several neighbouring plants that relied on their proximity to materials supplied by the American TNC.

Activity 2

Analyse the relative importance of political factors and economic factors causing the global credit crunch. Construct a two-column table that maps these causes.

Geographical impacts of the global credit crunch

The wealth of nations

The interconnected nature of the world economy meant that all nations were affected to some extent by the global credit crunch, many severely so as they entered a period of economic recession (Figure 9.4). According to the IMF, immediate government aid needed to support financial systems in the USA, UK and the Eurozone totalled US$9 trillion dollars (when debt guarantees for banks are included). In the first few months of the crisis, G7 nations agreed to use all available means to support 'systemically important' institutions and prevent their failure. The UK injected cash into Lloyds TSB and Royal Bank of Scotland (RBS), two of Britain's largest TNCs; RBS is one of the biggest financial service providers in the world.

Figure 9.4 Nations that were especially badly affected by the global credit crunch

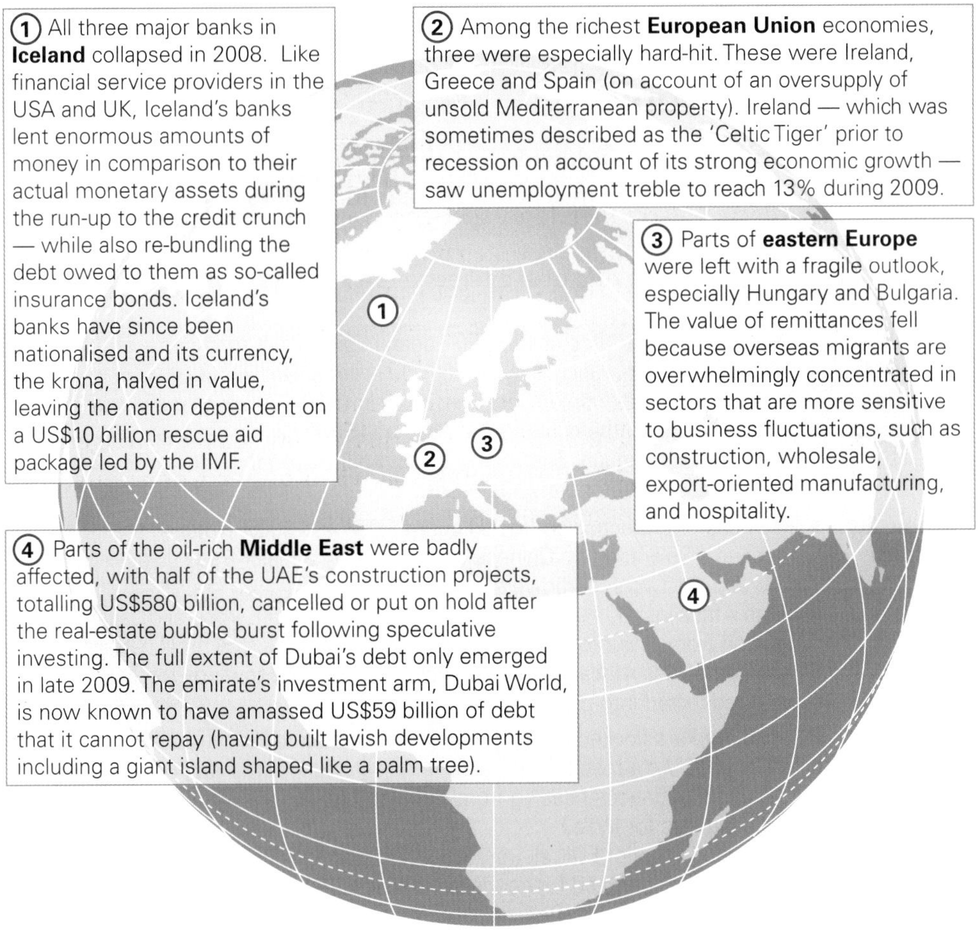

Most UK politicians viewed these banks as being too important to the UK's globalised and financially connected economy for their failure to be allowed by government; rescue packages for these beleaguered institutions have left the British public with an enormous collective debt potentially totalling US$1.4 trillion. As a consequence, the 2010s will be a decade of austerity in the UK, with higher taxes and sharp cutbacks in public spending on services such as higher education and healthcare. US taxpayers face a similar future, after the Senate supported key players such as Goldman Sachs to ensure they did not fail alongside Lehman Brothers and the 25 smaller American banks that filed for bankruptcy during 2008. American citizens must expect to see either public spending fall, or their taxes rise, or a combination of the

two. Total US national debt reached US$10 trillion in 2009, the highest ever recorded. Robert Wade, professor of political economy at the London School of Economics, observed that we now live in a world where 'the financial sector privatises profits but socialises losses'.

The wealth of people

Many millions of people worldwide lost their jobs during the global credit crunch. The US jobless total reached a 26-year high of 9% in 2009; Disney

Table 9.3 The changing nature of global interactions and flows in 2008–09

Migration flows and diaspora patterns	• Almost half of the UK's eastern European migrants returned home after the start of the credit crunch, leaving 700 000 still resident in the UK. • 20 million Chinese workers lost their work in export processing zones and returned home to the countryside; 9000 of 45 000 export factories in the Pearl River delta (Dongguan, Shenzhen, Guangzhou) closed their doors for a period, causing massive temporary job losses.
Commodity and trade flows	• Instances of protectionism tripled during 2008 and 2009. Notably, the USA imposed a 35% tariff on Chinese tyres, sparking a major trade dispute between the two superpowers. • South Africa lost 50% of its iron ore export trade with Europe and Japan; but demand from China for infrastructure projects rose by 35%. This pattern was repeated across Africa, as exports previously directed towards the EU and USA were redirected into China.
Financial flows	• Indian workers flocked home from Dubai as construction projects ground to a halt in 2008. As a result, financial remittances plummeted too. (With US$8.8 billion of remittances heading from Dubai to Kerala state alone in 2007, this was a big loss for India.) • Total world value of FDI fell from US$1.7 trillion in 2007 to US$1.2 trillion in 2008 as falling demand for goods and services led TNCs to suspend expansion plans for new markets. Only investment in agriculture went up, boosted by food security concerns and land grabs by emerging powers.
Information flows	• Information flow was one area that was unaffected. Increased membership of global virtual groups like Facebook and Twitter was a globalisation success story of 2008 and 2009. • Also in 2009, SEACOM completed its 17 000 km submarine fibre optic cable connection linking east Africa to global networks via India and Europe. • Chinese leaders continued to censor information flows into and out of China, building into a major commercial conflict with Google by the start of 2010.
Tourism flows	• Visitor numbers to Thailand fell by 20%. Many other countries recorded a fall-off in tourist numbers. • Internal tourism flows rose for many countries as people opted for a 'staycation' instead of a foreign vacation. In the UK, many more people stayed at home and took holidays locally in Scotland or Cornwall instead of heading overseas.

axed 1900 jobs at its theme parks, a clear symptom of a crisis of consumption in the US economy. Elsewhere, the World Bank estimated that 90 million people in Africa, Asia and Latin America were pushed back into extreme poverty by the crisis (meaning their per capita incomes fell below US$1.25 per day). Millennium Development Goals for 2015 have become harder to meet as a result of lower earnings and job losses experienced by some of the world's poorest people. International aid flows fell in value too, as rich nations, themselves in debt, tightened their purse strings.

At the 2005 Gleneagles summit, the G8 made a commitment to double aid to Africa by 2010; the events of 2008–09 ensured that only half that promise was met. The poorest countries face a long recovery now that aid, trade, remittances, tourism revenues and FDI have all fallen in value (Table 9.3). A lone beacon of hope for some African nations has been Chinese FDI, which, in stark contrast to other flows, has continued to grow in the aftermath of the global credit crunch.

At the opposite end of the income spectrum, some of the world's richest people watched as large parts of their fortunes vanished overnight. Of the 1125 dollar billionaires in 2008, only 793 retained that status a year later (though numbers have since recovered). Lakshmi Mittal, the Indian steel titan, lost US$26 billion of his fortune; Roman Abramovich's wealth slipped from US$23 billion to US$14 billion; and Microsoft's Bill Gates saw his US$40 billion estate halve in value. Major shareholders of several high-street chains — including Woolworths and Borders — also suffered major losses as these firms filed for bankruptcy.

Activity 3

What is the meaning of 'financial connectivity'? How did financial connectivity contribute to the global credit crunch?

The environment

Plunging world economic output after September 2008 led to a cut in global greenhouse gas (GHG) emissions far surpassing the results of any international climate treaty. Total GHG output declined by around 3% during 2009. Falling demand in all major markets for manufactured goods meant more industrial plants operating below full capacity and fewer new power stations being built. Many prestige construction projects were cancelled or put on hold, notably so in the global hub of Dubai. Airlines cancelled flights as individuals and families cut back on their holiday spending.

The global credit crunch achieved, albeit temporarily, something that no government leader has ever asked for as part of national or international efforts to tackle climate change: namely a fall in retail spending and corresponding shrinkage of consumer society's carbon footprint.

Globalisation's power shift

China emerged from the global economic crisis as its most significant winner, poised to become the world's second largest economy. While the American car giant General Motors was staving off bankruptcy during 2009, its Chinese operation was reporting record sales, up 50% on 2008. Despite some overseas success, General Motors cut its US workforce by 38%, axing 23 000 American jobs. During 2009, China:

- retained a national economic surplus of US$2 trillion while many major nations recorded unprecedented levels of national debt
- overtook Germany to become the world's largest exporter
- sustained GDP growth of 8% throughout 2009 while the UK and some other advanced economies entered recession (i.e. their GDP growth was negative for two successive quarters)
- overtook the USA to become the world's largest consumer market for vehicles
- increased its share of trade with other large emerging economies, including India, Indonesia, South Africa and Brazil
- advanced towards a free-trade agreement with Asian-Pacific states as part of a new China–ASEAN pact

India, China's superpower neighbour, also achieved impressive GDP growth of 6% in 2009. Both countries are members of the G20, which, despite being a relatively new grouping, is now superseding the longer-established G7/8 to become the more important locus for global geopolitical decision-making. The world's emerging economies, especially the 'BRIC' group of Brazil, Russia, India and China (or the variant 'BASIC' subset of Brazil, South Africa, India and China), are moving towards the centre-stage of world politics. At the Copenhagen climate change summit in December 2009, in what may prove to be a significant shift in the global balance of power, EU representatives had no real influence over decision-making by the emerging economies. Both India and China refused to commit to legally binding GHG reduction targets, although each made a domestic pledge to reduce the carbon-intensity of its rising economic output (Figure 9.5).

Figure 9.5 Attempting to reach a global agreement on climate change in Copenhagen, 2009

The Copenhagen Accord 2009

National governments met at the December 2009 climate change summit in Copenhagen. The aim of this summit was to provide the framework for a successor to the 1997 Kyoto Protocol, which asked all countries to commit to a firm target for GHG emission reduction.

Strengths of the Copenhagen Accord

- This was the most advanced global agreement on climate change yet and the furthest the world community has come in two decades of climate policy since the 1992 United Nations Framework Convention on Climate Change
- The most powerful key players reached a common agreement and all the signatories agreed to try to limit global warming to no more than 2°C above pre-industrial levels, which is vital according to IPCC* scientists
- Financial aid was agreed for poorer countries to fight climate change alongside a commitment by developed and emerging economies to reduce CO_2 emissions

* Intergovernmental Panel on Climate Change — the UN climate change body

Weaknesses of the Copenhagen Accord

- CO_2 emissions are still set to rise in China and India (as well as other emerging economies); both countries agreed to reduce the carbon-intensity of their GDP (which is growing very quickly) but not to cut their total emissions
- Several major countries felt ignored, including those of much of South America, notably Venezuela
- The EU aim to get more countries to set even higher targets for emissions reductions was not met and many people felt the decision-making was dominated by two superpowers — China and the USA

Pledges made by the global superpowers

European Union

The EU made a binding pledge to cut its emissions by 20% by 2020 (compared with 1990 levels), and also displayed a willingness to raise the target to 30% if other developed countries will agree to a much bigger commitment too (in line with advice taken from the IPCC).

USA

The USA is still the world's second biggest carbon dioxide emitter, producing 5900 million tonnes of CO_2 (e) each year. A bill mandating proposed cuts of 17% by 2020, compared with 2005 levels, had not yet passed through the US Senate at the start of 2010.

China

A non-binding target was set for 2020 to reduce the carbon intensity of China's growth. This will not lead to an absolute emissions cut but will significantly curb growth (meaning GHG emissions in 2020 will still be 40% higher than today, but lower than they might otherwise be).

India

A non-binding target has also been set by India to reduce the carbon intensity of its growth. A 24% reduction in emissions intensity is sought by 2024, equal to nearly 2000 million tonnes of CO_2 (e). India's rulers also believe developed nations should be doing more to help.

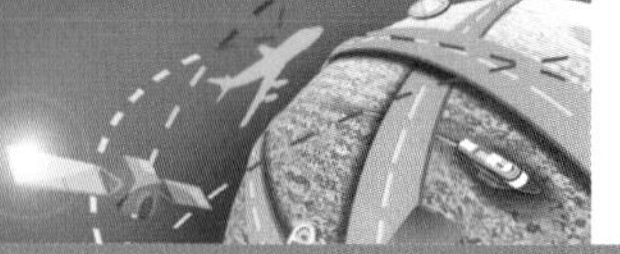

Table 9.4 Two views on how the global credit crunch affected the superpower balance between the USA and China

Signs of China's ascendancy	Signs that China is not yet the dominant superpower
• Following the credit crunch triggered by the sub-prime market collapse in the USA, Western financial ideologies and institutions have been left looking weak and discredited. As a result, leaders of African nations such as Nigeria may now be more likely to favour partnerships with Chinese TNCs, rather than with firms from Europe or North America. • China has become far more active in making energy deals with companies in politically sensitive parts of the world. For instance, Chinese companies are no longer cautious about investing in America's backyard of Latin America (PetroChina has made large investments in Venezuelan oil). In the Middle East too, China now has a presence — Iran is its third largest oil supplier (and there is no love lost between Iran and the USA). • China and the other large emerging economies are questioning assumptions about how Western organisations (including the IMF and World Bank) run global markets following the 2008–09 meltdown. • Cushioned by more than US$2 trillion of foreign currency reserves (the world's largest) and a massive trade surplus, China was buying up American debt at a rate of US$20 billion worth of US treasury bills a month in 2009. This is clearly symptomatic of a major power shift under way. • Right now, China is biding its time: the current global status quo is described by the Chinese as *yi chao duo qiang* — meaning there is one superpower (the USA) and several great powers (including China). Official policy up to now has been to not 'rock the boat' and to allow trade output to grow. Most recently, Chinese growth overtook that of the UK, Germany and Japan. Inevitably, China will one day be the world's number one superpower.	• China generates 8% of world domestic output, while the USA is responsible for 20%. So China's economy is still too small to become the world economy's 'locomotive'. The USA will remain the indisputable leader for some time to come; it currently maintains a powerful 17% share of the global vote on IMF decisions. This gap is just too large for a fundamental shift in America's relative position to take place anytime soon. • China needs to do much more to promote demand and consumption at home. Its export-dependent growth strategy will not allow it to move ahead of the USA for decades, even if demand for Chinese goods fully recovers. China rode the most recent wave of globalisation expertly but now needs a more diverse strategy. • For greater economic progress, China needs to encourage its citizens to spend more in shops rather than saving money for old age (introducing state pensions and health insurance could assist with this). Chinese people must save less and spend more to drive the economy forwards. • China also needs to shift to a form of growth that is less energy and commodity intensive. China's ecological footprint is huge and unsustainable. • With no democratic mandate, the unelected Chinese ruling elite still holds all the power. Twenty years after the Tiananmen Square massacre of anti-government protesters, the ruling communist party shows no signs of making a transition. If economic progress ever slackens, social unrest could grow, especially given public worries about corruption and state censorship (of the internet). • As one commentator has remarked, 'it may be premature to crack open the champagne. The Asian century is hardly preordained'.

Before the global credit crunch, Goldman Sachs forecast that China would overtake the USA in 2027 to become the world's number one economy and superpower. Post-credit crunch, many economists now believe the moment will arrive sooner, unless the Chinese economy overheats (as Japan's did in the early 1990s). There are two contrasting viewpoints to consider in relation to this assertion. One sees China on the verge of overtaking the USA to become the world's number one superpower; whereas the other does not (Table 9.4). However, in both scenarios China becomes less reliant on export income and more focused on building new sources of domestic demand for consumer goods and services in the future. There are already clear signs of this happening as wealth begins to spread or trickle down from core coastal regions towards the interior (Figure 9.6). Also, land reforms already under way could allow even the poorest farmers and farmworkers to gain access to state financial loans. The Chinese economy may continue to grow irrespective of whether the economic outlook for developed nations improves or deteriorates. China's sustained rates of high growth after the early 1990s took the world largely by surprise; bolstered further by its performance during the economic crisis, the Chinese 'miracle' appears far from having run its course.

Figure 9.6 China's continuing economic expansion during the global credit crunch (data show the annual percentage change in the first half of 2009)

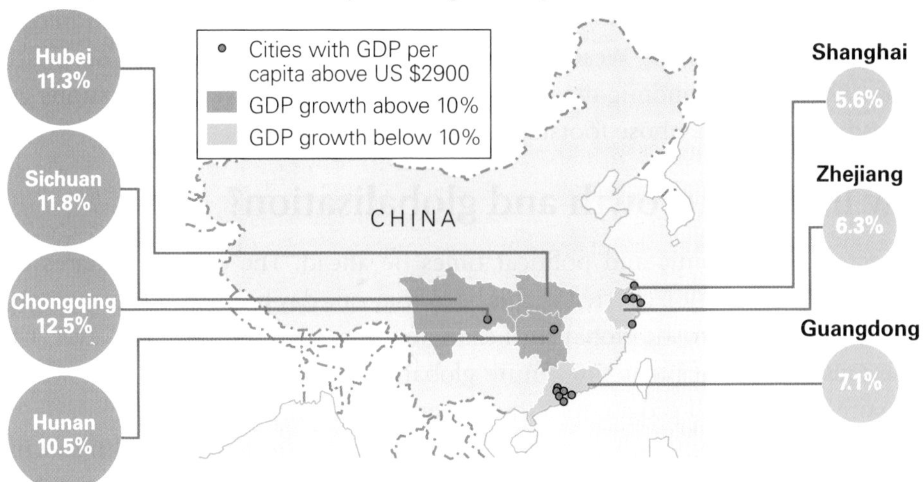

Global futures: business as usual?

The nations that have benefited most from globalisation until now have been those with sufficient geopolitical weight to control the terms of their own global interactions with other countries and with TNCs, in ways that yield significant economic and political rewards. For this reason, conventional wisdom has long been that globalisation and 'Americanisation' were one and the same thing. However, the picture is rapidly changing. One long-term effect of globalisation, only recently apparent, is the transfer of power to emerging economies which play by different economic rules from those favoured by the USA and other Western powers.

Globalisation prior to the global credit crunch was shaped by a **neoliberal** economic philosophy that promoted low taxation, privatisation and financial deregulation. It was the preferred model for policy-makers in Washington who, alongside major TNCs, can be described as the 'architects' of modern globalisation. Yet their system suffered catastrophic failings during the global credit crunch. As a result, new geographies of globalisation can be expected to emerge in the decades ahead, constructed increasingly by newly arrived global powers from developing nations. Their economic and social ideas may be very different from the ones to which we have grown accustomed.

For instance, China's citizens are advancing economically despite a continuing lack of democracy and personal freedoms; India and China reject key aspects of Western free-market economics (both countries enforce strict rules preventing full acquisition of any of their own big businesses by foreign TNCs). The new global superpowers are on uncharted development paths; they are not following the routes taken by Europe and North America. As a result, it is hard to see how 'business as usual', American-led globalisation can be resumed. For many Western intellectuals, all of this comes as a surprise: past predictive economic models have generally cast 'developed' nations as 'leading lights', in whose footsteps all others must follow.

New limits to growth and globalisation?

Interesting economic and political times lie ahead. The different phases of postwar globalisation charted in this book may one day be viewed as precursors to a far more dramatic global shift than anything seen hitherto. However, no serious attempt at predicting future globalisation trends can be made when many more critical geographical questions await an urgent answer:

- Will world population continue to grow rapidly from nearly 7 billion in 2010 to over 9 billion by mid-century before stabilising, as demographic models suggest? Meeting the basic resource needs of another 2 billion people will present a major global challenge.
- Can collective climate change mitigation efforts by the world community safely secure a maximum global temperature rise of 2 °C? This is identified by the majority of leading scientists as the critical safety threshold.

- Will existing local political and territorial disputes find reconciliation or could some escalate into larger regional conflicts? In particular, will friction between the Western powers and political Islamism subside?
- Will global water, food and land shortages bring increasing international conflict? Rising affluence in Asia and the Middle East means it may become harder to meet the changing demands of all nations.
- Can global society adapt to fossil fuel energy shortages once oil production has peaked and production levels fall? Humankind's hopes for the future remain pinned on innovation in the field of renewable energy.
- Are some current global health threats, such as a major flu pandemic, likely to escalate?

Figure 9.7 Globalisation's relationship with twenty-first-century challenges whose solutions require political cooperation at a global level

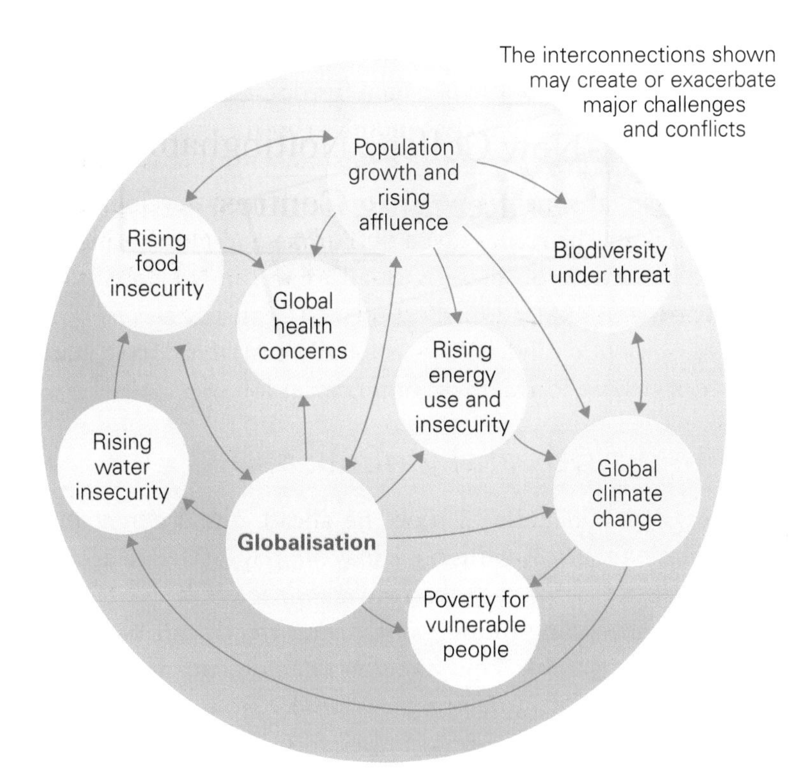

These issues are interconnected and none can be tackled in isolation from the wider set of concerns (Figure 9.7). Back in the 1970s, there was widespread belief that the world was on the verge of crisis due to population growth exceeding Earth's carrying capacity. This was known as the 'limits to growth'

hypothesis. Such pessimism gave way to greater optimism in the 1990s, as technologically assisted globalisation provided new grounds for excitement (especially for those in the driving seat). Now in the twenty-first century, we are experiencing new 'limits to globalisation'. The unstable nature of the system has been revealed and globalisation looks set to continue creating new challenges and tensions. The hope is that it may also provide the means to solve them.

Activity 4

Which twenty-first-century challenge do you see as being the greatest obstacle to sustained economic growth as a result of globalisation? Give reasons for your answer.